The Electron

by

Dennis Morris

Published by: Abane & Right

31/32 Long Row

Port Mulgrave

Saltburn

TS13 5LF

United Kingdom

01947 840707

dennis355@btinternet.com

May 2016

Contents

Contents

Properties of the electron

Electron mass $\qquad\qquad\qquad m_e = 9.109 \times 10^{-31} Kg$

Electron charge $\qquad\qquad\quad\ e = 1.602 \times 10^{-19} C$

Angular momentum $\qquad\qquad \dfrac{\hbar}{2} \quad$ up or down

Magnetic Dipole Moment (Bohr magneton)

$$\mu_B = \frac{e\hbar}{2m_e} = 9.274 \times 10^{-24} \ JT^{-1}$$

Lande g-factor $\qquad\qquad\qquad\ \sim 2$

Lepton number $\qquad\qquad\qquad 1$

Baryon number $\qquad\qquad\qquad 0$

Compton wavelength $\qquad\qquad \lambda_0 = \dfrac{2\pi\hbar}{m_e c} = 2.426 \times 10^{-12} M$

Bohr radius $\qquad\qquad\qquad\quad a_0 = \dfrac{4\pi\varepsilon_0 \hbar^2}{m_e e^2} = 5.929 \times 10^{-11} M$

The constant of proportionality of the magnetic dipole moment and the angular momentum is the gyromagnetic ratio, γ. We have:

$$\vec{\mu} = \gamma \vec{L} \qquad : \qquad \gamma = 2.$$

Planck's constant $\qquad\qquad\quad \hbar = 1.055 \times 10^{-34} Js$

Speed of light in vacuum $\qquad\ c = 2.998 \times 10^8 MS^{-1}$

$1 \ eV = 1.602 \times 10^{-19} J$

$1 \ \text{Å} = 10^{-10} M$

Introduction

This book is about the electron. We cannot tell the story of the electron without also telling the story of the electron's partner particle the neutrino. Thus, this book is about the electron and the neutrino.

People who have studied physics and chemistry at secondary school level or at undergraduate university level often have an understanding of the electron that is a century out of date. Modern quantum field theory sees the electron very differently from how physicists saw the electron in the 1920's or how Schrödinger and his peers saw the electron in the 1930's.

This book presents the modern QFT view of the electron and the neutrino to the reader.

Do not be daunted; things are much simpler than they sound.

Having said the above, this book does more than present only the modern QFT understanding of the electron and the neutrino. There are within modern QFT many aspects of our understanding of the electron that still discombobulate physicists. How can the electron occupy zero space? How can the electron be a wave and a particle at the same time? What is the nature of the intrinsic spin of an electron and why is electron spin directionally quantitised – either up or down but nothing in between? There are many other disconcerting aspects to the electron.

There are even more discomforting mysteries in our QFT understanding of the neutrino. Indeed, one of the central problems of modern particle physics is the 'neutrino mass problem'; it seems that neutrinos have both zero mass and non-zero mass at the same time.

As well as presenting the modern QFT view of the electron, this book presents answers to many of the unanswered questions about electrons and neutrinos. We come to these answers by approaching the electron and the neutrino from a different perspective to the conventional mantra. We walk *"... a road less travelled ..."* and seek a new perspective from which to view the electron and the neutrino.

In this book, we present the electron as a quaternion rather than as a pair of complex numbers. We similarly present the neutrino as a quaternion. We are very pleased to be able to say that using the quaternion approach gives answers to many of the discombobulating aspects of the conventional QFT understanding of electrons and neutrinos. In particular, we are able to rewrite the Dirac equation as a quaternion equation. The mathematics of the quaternion Dirac equation are much tighter than the mathematics of the conventional Dirac equation because we are within a division algebra. The quaternion Dirac equation leads to the view that there are no Majorana fermions, the electron field and the neutrino field are intimately entwined, and it seems to solve the neutrino mass problem and save the Standard model.

If walking *"... a road less travelled ..."* has solved the neutrino mass problem, then we have solved one of the most important problems of modern quantum field theory. The solution of the neutrino mass problem is that neutrinos are massless but the squared neutrino field is massive – quaternions can be like this. The details are given towards the end of this book.

We are also able to explain the directional quantitisation of electron spin – quaternion space is like this – and many other aspects of the nature of electron spin – quaternions are like this as well.

We hope, as well as walking *"... a road less travelled ..."*, the reader will gain a deep understanding of the conventional QFT view of the electron and the neutrino from this book. We have avoided the intense mathematics of QFT and preferred physical explanation over mathematical proof. We hope this policy has made this book into a readable and informative book without burying the reader under the mathematics.

Chapter 1

More Introduction and Historical Details

Atomic matter:

From as far back in history as we have records, humankind has questioned the nature of matter. Perhaps the most famous of such ancient questioners was the pre-Socratic ancient Greek Democritus (460 BC-370 BC) who is often referred to as the father of science[1]. Democritus is credited, perhaps most unfairly to many others, particularly Leucippus and Miletus, with the idea that matter is comprised of tiny little bits rather than being continuous; we say matter is atomic. Democritus is also associated, more importantly many would say, with the idea that every natural event is the consequence of natural laws rather than the intervention of the gods.

In the time of Democritus, the idea that matter is a collection of tiny atoms rather than a continuous medium was no more than assertion. Although Democritus could continually sub-divide a lump of butter into smaller and smaller amounts, the ancient Greeks simply did not have the technology to go all the way down to a single molecule of butter, divide that molecule, and then show that neither of the divided parts of the butter molecule was butter.

The idea that matter was divided into atoms remained no more than assertion for more than two millennia until chemists began to realise than certain elements always combine in particular ratios to form compounds and that this might be because the amounts of elements in the compounds were integer numbers of atoms. Building upon such work, Dmitri Ivanovich Mendeleyev (1834-1907) produced the periodic table of the chemical elements in which the naturally

---

[1] Thales circa 500 BC has also been referred to as the father of science.

occurring chemical elements are classified[2] in a regular order. When Mendeleyev produced his periodic table, not all the 92 naturally occurring elements were known, but, none-the-less, the regularity of the chemical properties of the elements was apparent.

Since there are 92 different naturally occurring chemical elements, and since the chemical properties of these 92 elements repeat regularly, it is not foolish to question whether the atoms of these different chemical elements might each be comprised of varying numbers of sub-atomic particles in some regularly repeating way. Of course, we now know that the different chemical elements, their isotopes, and their ions are comprised of different numbers of protons, neutrons and electrons.

Aside: Mendeleyev habitually had a haircut only once each year in springtime when he called upon a local shepherd to shear him.[3]

Sub-atomic matter:

In 1896, J. J. Thompson (1856–1957), John Sealy Townsend (1868-1957), and Harold A. Wilson (1874–1964) discovered the first sub-atomic particle when they discovered the electron. Subsequently, other sub-atomic particles like the proton and the neutron were discovered, and, by the early 20th century, humankind had come to the understanding of atomic structure which is taught to school children throughout the world today and which explains Mendeleyev's periodic table of the chemical elements[4].

Although J. J. Thompson is often, unfairly to his colleagues, credited with the discovery of the electron, the concept of the electron predates Thompson's discovery. Previous assertions of the existence of the electron can be found from Richard Laming (1798–1879),

---

[2] Mendeleyev's periodic table omitted several elements which were unknown at the time. The classification scheme effectively predicted the existence of these undiscovered elements.

[3] The history of science is replete with characters whose nature is beyond the imagination of any fiction writer.

[4] Not quite a complete explanation; we need the Pauli exclusion principle also.

William Weber (1804–1891), George Johnson Stoney (1826–1911), and Hermann von Helmholtz (1821–1894). Indeed, it is from the work on electrolysis done by Stoney in 1874 that the name electron, electric-ion, is derived.

Building upon Thompson's discovery of the electron, Robert Millikan (1868–1953) and Harvey Fletcher (1884–1981) determined the amount of the basic electric charge of the electron in their famous 1909 oil drop[5] experiment[6].

No hard balls:

Dating all the way back to Democritus, we have the idea that matter is comprised of particles most often envisaged as tiny hard balls. Today, we know that Democritus understood it wrongly. The electron, and all other sub-atomic 'particles' are not particles anything like tiny hard balls. Sub-atomic particles have no size at all.

We really ought to take pause and think about the previous sentence. How can anything occupy zero space? How can anything have zero spatial extent?[7]

When we were still babies in our cots, we were taught that an atom of, say, silver was comprised of a central nucleus of hard balls stuck together and a number of electrons (tiny hard balls) orbiting around that nucleus in a way very similar to the way in which planets orbit around the sun. If we asked our teachers, they would confirm that most of an atom is empty space and they might give us an analogy that pictured the central nucleus to be the size of a pea and pictured the electrons to be orbiting around this 'pea' perhaps twenty miles distant from it. Between the nucleus and the electrons was nothing but empty space. We might have been told that an atom is 99.99%

---

[5] Millikan R. A. (1913) On the elementary electric charge and the Avogadro constant. Phys. Rev. 2 (2) 109-143.

[6] We now know that Millikan and Fletcher 'rigged' the results of this experiment a little, but the experiment is essentially correct.

[7] The number one has zero spatial extent; as does the English language, but this does not seem to help.

empty space. We might even have been told that an electron is a point particle that has no size and therefore occupies no space.

Aside: It is possible to excite atoms in such a way that the outermost electron(s) are very distant from the nucleus; atomic diameters of $\sim 1\mu m$ have been observed. Such atoms are known as Rydberg atoms.

By the time we arrived at infant school, we were given atom-smashing machines to play with and we began to ask of what was the nucleus comprised. We were told that the atomic nucleus is comprised of neutrons and protons and that it has a definite measurable spatial size. We were told that experiments associate neutrons and protons with a definite size of circa 10^{-15} metres. This is the spatial extent of the electric and magnetic fields of these particles[8].

We now know that protons and neutrons are comprised of the point particles we call quarks and gluons and that 'inside' the proton and the neutron is just more empty space. Eventually, we come to realise that an atom is not just 99.99% empty space; an atom is 100% empty space. There is not a tiny hard ball anywhere in sight.

Today, we still talk of electrons, protons, mesons, and quarks as sub-atomic particles. Indeed, we even call the study of these entities 'particle physics'. We ought more properly to call the study of electrons and protons 'empty space physics'.

Then there were waves:
Things became even less like Democritus had envisioned them when, in 1924, the French physicist Louis de Broglie, in his paper *Recherches sur la theorie des quanta*[9], proposed that an electron has wave-like properties. We all know, or thought we knew, that tiny hard balls do not wave up and down as they move from place to place,

---

[8] Neutrons have zero electric charge, but they still have a magnetic moment 'caused' by the electrically charged quarks which constitute them.
[9] Louis de Broglie Thesis Paris Ann. De Physique (10) 3 22 (1925)

but it seems that electrons do have these wave like properties. Only a few paragraphs ago, we recounted how our infant school education had led us to see electrons as point particles. We now recount how, when we arrived at primary school, we were told that electrons are waves. If we asked what is waving or in what medium the waves are transmitted, we were told that probability is waving[10] and that there is no medium.

If we asked about the spatial extent of these electron waves, we were told that electron waves do not exist at a particular point in space. Electron waves are at least sufficiently spatially extensive to pass through two spatially separated slits simultaneously and thus interfere with themselves, as do classical waves. Thus, we have it that electrons are both point particles and spatially extensive waves. It is no wonder that physics students spend so much time in the pub.

Then things became complicated:

While at primary school, we were further taught of things like the Pauli exclusion principle and intrinsic spin. Apparently, the point particle that is the electron has angular momentum of some special type called intrinsic spin. How can an object of no spatial extent be spinning?

Things became even more obscure when we began our secondary school studies and found that the electron is something to do with a neutrino, and, when an electron is massless, which it never is, an electron is a neutrino some of the time but only if it is left-handed. Right-handed massless electrons are never neutrinos. Of course, neither left-handed nor right-handed massless electrons exist in our 4-dimensional space-time, but the concept of them underpins electroweak theory. This might seem confusing to the reader; if it does seem confusing to the reader, it is because the reader is of sufficient wit to recognise confusion when she sees it. Well done! You have clearly understood that science is confused about the electron. We will write extensively of these matters latter in this

---

[10] Was your author ever confused enough to swallow this nonsense? Yes.

book; until then, the reader is advised not to worry about this confusion.

When I was a baby, I understood the nature of the electron exactly. After being educated, I no longer understand of the nature of the electron.

We just do not understand the electron:

This book is about the electron. Your author would like to be able to claim honestly that he completely understands the nature of the electron and so he is qualified to write a book about the electron. Sadly, your author cannot honestly claim to understand the electron. His only comfort in this soul crushing intellectual desolation is that he is not the only person on the Earth who does not understand the electron.

We know that the electron is not a tiny hard ball. We know the electron takes up zero space. We know the electron is a spatially extensive wave. We know that all electrons are identical – this in itself is amazing. We know that electrons all have electric charge, but we do not really understand the nature of electric charge. Just what is electric charge?

We have very successful theories that unify electrons with other particles like photons and weak force bosons, and yet, if the truth was told, in spite of the success of these theories, we do not in our hearts understand what is happening in the sub-atomic realm.

Over the last century, the acceptance of our impotence in this regard has changed the way physicists view themselves. A century ago, physicists believed they would one day know and properly understand everything. Einstein referred to 'The Grand Unified Field Theory', which is a single theory of everything; he spent his life searching for 'The Grand Unified Field Theory'. Today, physicists claim they can do no more than construct a model of reality and that, in spite of the success of this model, reality is beyond human understanding. Your author is more optimistic. Your author opines that we are on the verge of breaking through the fog of conceptual confusion and that we will one day understand everything.

In this book, we will present to the reader many aspects of the electron, and we will present some mathematics that seems to be relevant to an understanding of the electron. We will also present material that gives a view of the electron that is different from the conventional view; we will walk *"...a road less travelled..."* and see to where this road less travelled leads. We will assert that our presentation is as good as the presentation of any other person. However, unfortunately, even walking down the *"...road less travelled..."*, when all is said and done, we are unable to explain, as one might explain to a twelve year old, the true nature of the humble electron. The reader will have to form her own understanding of the electron. There is still work to be done.

An aside on the modern education system:
Reading the above, the reader might be struck with how much better the education system was in the 1960's than it is today. It is generally agreed that the literacy and numeracy levels of the younger generation within western societies are worse than they was[11] in the 1960's.

---

[11] Even in the 1960's literacy was not perfect.

Chapter 2

How Electrons Move

This short chapter is not really in sensible presentational order. However, your author feels that to introduce this material now will help the reader to understand the following chapters. We present the current understanding of how electrons move through our 4-dimensional space-time without explanation because we have not yet developed our knowledge sufficiently to understand the explanation – that's why this chapter is out of order.

This understanding of electron motion through our 4-dimensional space-time is the understanding afforded to us by quantum field theory, QFT; it is very different from our classical understanding of electron motion. The presentation is incomplete, but we present the general view. These matters will be approached in more detail later in this book.

A single electron:

In the view of QFT, as an electron exists, it emits and absorbs photons of light. This is what electrons do. This is all electrons do. They simply sit there emitting and absorbing photons of light[12]. Every electron emits and absorbs trillions of photons every second. This is very different from the classical view of an electron.

Each time an electron absorbs or emits a photon, because photons have momentum, the electron is 'pushed' or 'pulled' by the momentum of the photon in one direction or the other. However, electrons emit and absorb photons very symmetrically with the effect that a photon 'push' in one direction is met, perhaps slightly later, by

---

[12] There are complications like neutrinos to this simple picture, but, for now, we will be content with this simple picture. The photons are virtual photons – more later.

a photon 'pull' in the reverse direction. The result is that the electron 'jiggles about' microscopically but essential stays still macroscopically.

If the electron is moving in a straight line through our 4-dimensional space-time, it will macroscopically move as if it were a particle with a definite momentum moving in a definite direction. Microscopically, the electron will 'zig-zag' all over the place as photons are emitted and absorbed, but, on average, the directions and amounts of momentum of the emission and absorption will balance and the electron will move macroscopically on a straight line. We have a picture:

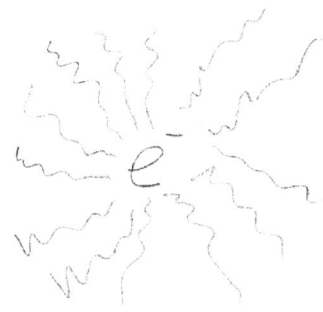

The technical aspect of this understanding is that the probability amplitude[13] for an electron to emit or absorb a photon of a given momentum to the left is equal to the probability amplitude for an electron to emit or absorb a photon of a given momentum to the right; the balance of probability amplitudes is what keeps the electron moving in a straight line. Another picture:

---

[13] Probability amplitude is a technical term which just means probability. We could have omitted the 'amplitude' word.

The reader might be struck by this excellent picture of an electron; it is nothing like a hard ball.

If a moving electron enters an electric or a magnetic field, then the path of the electron is 'curved' by the electric or magnetic field. The nature of the electro-magnetic field is that it influences the emission and absorption of photons by the electron in an unbalanced way. This imbalance in photon emission and absorption by the electron is what changes the path of the electron through our 4-dimensional space-time. The imbalance in the momentum of the emitted and absorbed photons 'pushes the electron off course'.

We see that an electron is far from being a particle that just sits quietly where it finds itself. The electron is very active and always interacting with its own electromagnetic field by emitting and absorbing photons. By the way, the photons are virtual photons - more explanation will be given in later chapters.

We have made no mention of electron waves. The above applies to only electron particles.

The speed of light:
The conventional view is that there is a probability amplitude that a photon of light will move slower than the speed of light and that there is a probability amplitude that a photon of light will move faster than the speed of light. These two probability amplitudes are equal, and so, macroscopically, there is a balance and a photon of light moves at the speed of light. Your author questions this view. Your author opines a photon moves at the speed of light because it is outside of our 4-dimensional space-time and that anything outside of our 4-dimensional space-time moves infinitely fast but that we perceive it to move at the speed of light; this is not the conventional view.

The conventional view is that there is a probability amplitude that a photon of light will move to the left rather than fly straight ahead and there is a probability amplitude that a photon of light will move to the right rather than fly straight ahead. These two probability amplitudes are equal, and so, macroscopically, there is a balance and a photon of light moves in a straight line. Of course, the actual direction of the straight line is determined by the curvature of our 4-dimensional space-time as presented in the general theory of relativity[14].

We see that quantum field theory explains why light moves in a straight line at a constant speed, but it needs light to be particles rather than waves for the explanation to work. Of course, we accept that light is particles, photons, because this explains so much of the world.

Hidden assumption:

There is a hidden assumption in this entire chapter. We have assumed that the electron 'knows' enough about the directions and structure of our 4-dimensional space-time to be able to emit and absorb photons equally in all directions. Not all types of empty space have the same structure.

---

[14] See Dennis Morris Upon General Relativity

Chapter 3

Cathode Ray Tubes

In 1996, televisions were cathode ray tubes, and most homes had one. A hundred years earlier, in 1896, the cathode ray tube was some of the most advanced laboratory equipment available. The cathode ray tube is so called because it is a tube that has cathode rays within it. We have a picture of one:

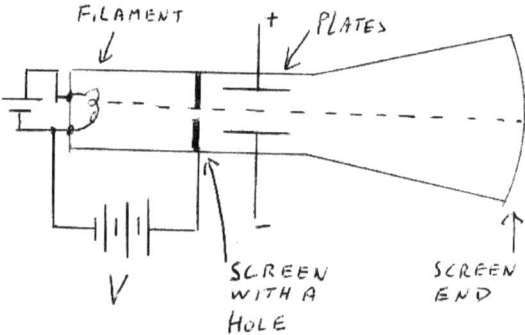

The cathode ray tube is evacuated to remove the air molecules – they get in the way of the cathode rays[15]. When a voltage is applied across the filament (cathode), electrons 'boil' from the filament. These electrons are accelerated along the cathode ray tube by the electrical potential difference between the screen with a hole and the filament. Many of the electrons hit the screen with a hole, but some pass through the hole forming a beam of collimated electrons; posh cathode ray tubes might have two such screens for better collimation. This collimated stream of electrons is the cathode rays.

---

[15] At atmospheric pressure, electrons have free paths of approximately $10^{-6} M$ before hitting an air atom. At 10^{-8} atmospheres, the electron's free path is about 2 metres.

The charged plates form an electric field. There are also current carrying coils (not shown[16]) with the plates that form a magnetic field perpendicular to the electric field. When the beam of electrons hits the fluorescent screen at the end of the tube, the spot where they hit the screen fluoresces (glows). The electric field will deflect the electron beam, cathode ray, thereby deflecting the position of the glowing spot on the fluorescent screen, and the magnetic field can be adjusted in strength to undo that deflection.

In 1896, neither Thompson nor anyone else knew about electrons. To Thompson and his colleagues, there were mysterious electrically charged rays, called cathode rays because they emanated from the cathode filament, moving down the length of the tube. Thompson was able to measure the ratio of the mass, m, to the charge, q, of the cathode rays. He repeated this experiment using several different metals for the cathode filament and measured the ratio of the mass to the charge of the cathode rays from each element. The mass to charge ratio was the same for all the elements that he used as the filament. Thompson then declared that the cathode rays were streams of charged particles, which he called corpuscles, and that these corpuscles were a basic constituent of all types of matter (of all different elements). He further concluded that each corpuscle was less than a thousandth of the mass of the lightest chemical element hydrogen. Today, we call these corpuscles electrons; an electron is $\dfrac{1}{1836}$ the mass of a hydrogen atom.

This is how electrons were first discovered. Well, this is how the concept of the electron first came into human understanding. Arguably, the first person to be electrocuted was the true discoverer of the electron.

---

[16] Clearly, your author is an artist of considerable skill, but coils were just too much for him.

Single electrons:

Now, suppose we lower the voltage in the filament of the cathode ray tube so that only one electron per second is emitted in the direction of the hole in the first screen. What do we observe at the fluorescent screen? We see individual flashes one second apart at the same point on the fluorescent screen as each electron hits the screen at the same point. Clearly, our electrons are particles or at least have particle like properties. The electrons seem to be following a definite straight trajectory through our 4-dimensional space-time from the filament to the fluorescent screen. We can bend this trajectory using the electric plates or the magnetic coils. Everything works as if the electron was a particle moving in a definite trajectory.

From our observations of the cathode ray tube with a screen that has a single hole, we conclude:

i)   Electrons behave like individual particles.
ii)   Electrons follow a definite straight trajectory through our 4-dimensional space-time.

Aside: The mass of an electron, or the mass of anything, can be measured in two ways. One way is by using $F = ma$ and asking how much force is needed to accelerate the electron by a given amount. The other way is by using $E = mc^2$ and asking how much energy is produced when an electron is annihilated (by collision with a positron).

Two holes in the screen:

Suppose we use in our cathode ray tube a screen with two holes very close together rather than only one hole. Such an experiment with two holes has never been done, but we know the result because experiments have been done using crystals and other materials that effectively are a screen with two small holes close to each other. The most famous of these experiments were the experiments of Clinton

Davisson (1881–1958) and Lester Germer (1896–1971) done between 1923 and 1927[17].

The electron now has two possible trajectories that it might take from the filament to the fluorescent screen. Since the electron is a particle, it will, according to our naive understanding of the universe, pass through either one hole in the screen or the other hole in the screen. Remarkably, the electron passes simultaneously through both holes. It takes both trajectories available to it. How do we know this? We know the electron passes through both holes at the same time because it interferes with itself and it can do this only if it passes through both holes simultaneously.

If we send many electrons towards a single holed screen, we get a single spot on the fluorescent screen. If we send a wave towards a single holed screen, we get a single wave emerging from the hole. If we send a wave towards the two-holed screen, bits of the wave pass through each of the two holes and, upon emerging from the two holes, spread out until the two bits of the wave meet and interfere with each other forming interference fringes, alternating bands of brightness and darkness, on the fluorescent screen. If we send many electrons towards the two-holed screen, we observe interference fringes on the fluorescent screen. We do not see these if we have only one hole in the screen.

It is as if the electrons form a wave of electric charge and electron mass that passes down the cathode ray tube through the two-holed screen. It is as if the electron is a particle when only one trajectory is available to it but is a wave when more than one trajectory is available to it. This dual nature of electrons, and all atomic particles, is called wave-particle duality.

Single electrons:
Suppose we send only one electron at a time towards the two-holed screen. The single electron hits the fluorescent screen at one

_____

[17] Davisson C, J; Germer, L, H (1928-04001) 'Reflections of electrons by a crystal of nickel' Proceedings of the National Academy of Sciences of the United States of America 14(4) 317-322

particular point; it is not spread out like a wave into interference fringes. However, where on the fluorescent screen the electron hits is a matter of probability. The electron hits the fluorescent screen at a point different from where electrons passing through a single holed screen hit the fluorescent screen. Suppose we send many single electrons one at a time down the tube with a two-holed screen and we record the positions where each solitary electron hits the fluorescent screen. After several hundred solitary electrons have passed down the tube, we find that the electrons have hit the fluorescent screen at different points. However, these different points of impact are not without order; we have interference fringes on the fluorescent screen. Clearly, even solitary electrons are like waves when offered more than one possible trajectory to follow.

Above, we have said that electrons are point-like particles of zero spatial size. If a single electron can pass through two spatially separated holes at the same time, then it must be of some spatial size. What is happening here?

An equally important point is that, for the individual electron passing through a two-holed screen, the position at which that electron hits the fluorescent screen is undetermined. Exactly the same series of events leads to different outcomes for each electron. A hundred years ago, physicists were certain the universe was deterministic and that a set series of events would always result in the same outcome. According to determinism, the electrons should hit the fluorescent screen at the same place every time. The electrons do hit the fluorescent screen at the same point every time if we use a single holed collimating screen, but if we have a collimating screen with two holes, the electrons do not hit the fluorescent screen at the same point. Determinism is dead, and we have a random outcome to a set series of events. Probability plays a part in physics.

Probability playing a part in physics? Is your author insane? Experts have said not.

Aside: The reader might have read elsewhere that the electron 'decides' whether to be a wave or a particle when an experimenter changes the observing apparatus from a single holed screen to a two-holed screen. Electrons do not decide; electrons just are. An electron has wave like properties when offered more than one trajectory to follow, and an electron has particle like properties when offered only one trajectory to follow. An observer sees that which they contrive to see by the arrangement of the observational apparatus.

i)   Electrons behave like waves when offered more than one trajectory to follow.

ii) Electrons behave like particles when offered a single trajectory to follow.

iii) The outcome of events is not deterministic. Probability plays a role in physics.

We might take the view that an electron takes every possible trajectory offered to it through our 4-dimensional space-time. If more than one trajectory is offered, wave-like phenomena will automatically result. This does not mean that the electron is a wave; it means that the electron is travelling by more than one trajectory from its start point to its end point. You see, we are thus rid of having to think of an electron as a wave if we allow it to traverse more than one trajectory at a time.

Perhaps the previous paragraph was of such importance that it is worth repeating. For a century, physicists have thought of the electron as being both a particle and a wave. If we allow the electron to traverse two or more trajectories simultaneously, then we can take the electron to be a particle. We do not need the electron to be a wave if we allow that it traverses more than one trajectory at a time.

The reader should think about the previous paragraph. Why is it that a cricket ball flies from the bat in only one trajectory? We are so accustomed to this single trajectory behaviour that even to question it seems lunacy. Electrons force us to question the single trajectory concept. Unfortunately, your author has no idea why our 4-dimensional space-time is such that objects like cricket balls fly in only one direction.

Now the weird stuff:

How about no screen at all? Richard Feynman (1918-1988) used to put it like this. Imagine a screen with three holes in it rather than two; the electron must pass through all three holes. How about four holes or four million holes? Suppose we make the screen infinitely large and drill an infinite number of holes through the screen so that the screen no longer exists; the electron passes through every one of this infinite number of holes in the infinitely large screen.

Well, within a cathode ray tube, we might expect the 'no screen at all' to be no bigger than the diameter of the cathode ray tube, but there would still be an infinite number of holes in this 'no screen at all' screen.

The electron goes by every possible trajectory from the filament to the fluorescent screen. If unrestricted by the sides of the tube, does this include trajectories via distant galaxies? Yes, in defiance of the limiting velocity of our 4-dimensional space-time, it seems that every possible trajectory includes those trajectories that require the electron to exceed the speed of light. Nothing can exceed the speed of light within our 4-dimensional space-time; if Feynman's scenario is correct, the electron cannot be within our 4-dimensional space-time.

One of the basic underlying ideas of quantum field theory, QFT, is that particles like the electron do not travel through our 4-dimensional space-time along definite trajectories between two definite points at which they are observed if there is more than one trajectory available to them between the two points. Within QFT, this 'no definite trajectory' is more than just a concept; something very analogous to it underpins the mathematics of Feynman diagrams which are the principle tool in QFT used to calculate the outcomes of experiments – scattering amplitudes, radioactive decay and such like. In other words, the outcomes of experiments can be correctly calculated using mathematics based upon the 'no definite trajectory' concept. Within QFT, all possible trajectories are considered and added by using a path integral.

Indefinite trajectory in the time sense:

There is more. We said above that, when single electrons pass through a two holed screen, they hit the fluorescent screen at definite but different points and that the different points add together to make interference fringes on the fluorescent screen. Now, imagine the fluorescent screen was shaped in such a way that some of the points hit by electrons were at a distance from the filament different from the points hit by other electrons. Imagine a central projection of the fluorescent screen closer to the filament than the rest of the fluorescent screen.

We live in a 4-dimensional space-time in which the time direction is equivalent to a space direction – that is why we can rotate in space-time (change velocity). If the electron has no definite trajectory spatially, it must have no definite trajectory temporally. This is the basic equivalence of space and time in the theory of special relativity.

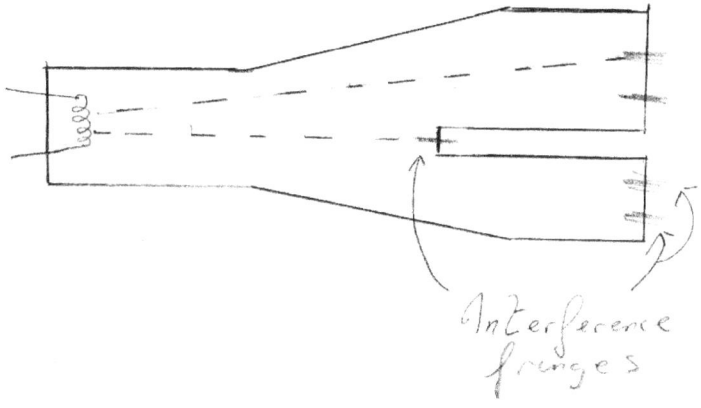

Interference fringes

If we see the electron as travelling down the length of the cathode ray tube, then, since some electrons will hit the fluorescent screen above or below of the central projection in our diagram, such electrons have 'decided' not to hit the central part of the fluorescent screen by the time they pass the central projection. However, this is a definite decision; this cannot be if the trajectories of the electrons are indefinite unless the time at which the electron hits the screen is also indefinite. The current understanding is that the trajectory of the electrons through 4-dimensional space-time is indefinite both spatially and temporally. It is as if, when more than one trajectory is available to the electron, time has no meaning for the electron just as spatial position in our 4-dimensional space-time (trajectory) has no meaning for the electron.

Aside: It is quite a thought to realise that if the space and time of the universe were two separate spaces, as Isaac Newton (1643–1727) thought them to be, then we could not change velocity (rotate in a 2-dimensional space-time plane) within our universe.[18]

The size of the electron:

Suppose we make the holes in the screen very small. Will we ever get to the point where an electron is too big to go through the holes? No is the answer. Using crystals with extremely small gaps between the atoms of their lattices in cathode ray tubes and other general scattering experiments have convinced physicists that electrons are point particles with zero spatial extent. Since we live in a single 4-dimensional space-time, we must surely treat time as we treat space. If the electron has zero spatial extent, then does it have zero temporal extent - exist only instantaneously? We will offer an explanation of this later when we consider quaternion space[19].

Of course, if an electron has zero spatial extent, where do we store its mass and its electric charge?

Photons and other particles:

The electron is not alone in having both particle like properties and wave like properties. Isaac Newton was of the view that light was a stream of particles, corpuscles. Later scientists formed the opinion the light was a wave until in 1905 Albert Einstein (1879–1955)

---

[18] See Dennis Morris : Empty Space is Amazing Stuff – The Theory of Special Relativity

[19] Rather than cruelly keep the reader in suspense, we look ahead. It seems that the electron can have spatial extent (go through two holes at the same time) only instantaneously and that it can have duration only if it has zero spatial extent.

explained the photoelectric effect by assuming that a beam of light was a stream of particles. We now have the view that light has both particle like properties and wave like properties. We call the particles of light photons.

Of course, the reader might well be aware that all sub-atomic particles, protons, neutrons etc. have both particle like properties and wave like properties. Indeed, even macroscopic objects like a virus have been shown to have both particle like properties and wave like properties. Given that a virus has been shown to have both particle like properties and wave like properties, we presume all macroscopic objects, like cricket balls or even planets, have both particle like properties and wave like properties.

Summary:

We have seen that there are 'things' which we call electrons which have a specific amount of electric charge and a specific amount of mass. Each electron has exactly the same amount of electric charge and exactly the same amount of mass. All electrons are identical.

Given two or more different trajectories through 4-dimensional space-time, these electron 'things' interfere in a wave like way not only with other electrons but, if alone, with themselves. These electron 'things' also appear at individual positions in our 4-dimensional space-time.

Electrons seem not to move through our 4-dimensional space-time on a definite trajectory unless only one trajectory is available to them. This indefiniteness is both spatial and temporal.

Electrons seem to have zero spatial size.

For electrons, we have:

1) All electrons have a definite mass
2) All electrons have a definite electric charge
3) All electrons are identical
4) All electrons seem wave-like

5) All electrons seem particle-like
6) All electrons seem to have zero spatial size
7) All electrons are of sufficient spatial extent to pass through two spatially separated holes in a screen simultaneously

Chapter 4

The Compton Effect

It is really good to know that electrons can have no definite trajectory through our 4-dimensional space-time unless there is only one trajectory available to them. However, even if they move along several trajectories, they do seem to retain a 'memory' of their previous momentum and their previous position in our 4-dimensional space-time. An electron appears to follow a definite trajectory through our 4-dimensional space-time because, each time it interacts in our 4-dimensional space-time, it 'remembers' the details of its previous interactions. We do not know how the electron 'remembers' the details of its previous position in our 4-dimensional space-time.

In 1923, Arthur Holly Compton (1892-1962) conducted an experiment[20] that later won him the 1927 Nobel prize. Without going into the details, Compton hit electrons with photons (x-ray photons) and thereby caused a change in the momentum of both the electrons and the photons in accordance with the conservation of momentum. The momentum of a photon is given by $p = \dfrac{h}{\lambda}$. The change of momentum of the photon was manifest as a change of the photon's frequency because the mass of a massless photon and the speed of a photon (the speed of light) cannot change. The effect is known as the inelastic scattering of a photon by a charged particle.

The Compton effect is taken as proof of the particle nature of light; it is also proof of the particle nature of electrons. It is also proof that electrons absorb (are hit by) and emit photons.

---

[20] Compton, Arthur, H. (May 1923) "A quantum theory of the scattering of x-rays by light elements" Phys. Rev. 21 (5) 483-502

Chapter 5

Magnetic Moment and the Stern-Gerlach Experiment

Consider a spinning sphere similar to the spinning Earth. Now, imagine that the spinning sphere is electrically charged. Moving electrical charge is associated with a magnetic field, and so rotating electrical charge is associated with a magnetic field. The magnetic field of such rotating electrical charge is a dipole represented by a vector pointing in the opposite direction to the vector representing the angular momentum[21], and the magnetic field is referred to as a magnetic dipole moment.

If we think of the electron as a classical spinning charged body, the rotating electrical charge generates a classical magnetic dipole moment given by:

$$\vec{\mu}_{classical} = \frac{-e}{2m_e} \vec{L} \qquad (5.1)$$

Wherein $\vec{\mu}_{classical}$ is the classical magnetic dipole moment vector, m_e is the mass of the electron, e is the charge of the electron, and $\vec{L}$ is the electron angular momentum vector. The classical angular momentum measures the rotational velocity of the electrical charge, and, classically, the classical magnetic dipole moment varies with the classical angular momentum.

The experimentally measured magnetic dipole moment of the electron is:

---

[21] The magnetic dipole moment vector points in the opposite direction to the angular momentum using our arbitrary definitions of current etc. Effectively, ignoring the signs of our arbitrary definitions, the magnetic dipole moment and the angular momentum vector point in the same direction.

$$\vec{\mu}_{electron\ actual} = -g\frac{e}{2m_e}\vec{L} \qquad (5.2)$$

Wherein g is a dimensionless number known as the g-factor or the Lande g-factor and $\vec{L}$ is the classically assumed angular momentum of the electron. The electron magnetic dipole moment has been measured to be $-9284.764 \times 10^{-27} J.T^{-1}$ to an accuracy of 7.6 parts in 10^{-13}.[22]

The amount of magnetic dipole moment associated with an electron is called a Bohr magneton named after Niels Bohr (1885-1962). The Bohr magneton is defined as:

$$\mu_B = \frac{e\hbar}{2m_e} \qquad (5.3)$$

This gives the electron's magnetic moment as:

$$\vec{\mu}_{electron\ actual} = -g\mu_B \frac{\vec{L}}{\hbar} \qquad (5.4)$$

Magnetic dipole moment is not quantitised. There are magnetic dipole moments of any value including values much less than the value associated with the electron. However angular momentum is quantitised. A quantum of classical angular momentum is an $\hbar$'s worth of angular momentum.

If we follow the classical picture and refer to the angular momentum of the spinning electron as spin, $\vec{S}$, or intrinsic spin to give it its Sunday name, then the magnetic moment of the electron becomes:

$$\vec{\mu}_{electron\ actual} = -g_s\mu_B \frac{\vec{S}}{\hbar} \qquad (5.5)$$

Classically, we would expect the spin g-factor, g_s, to be unity. The electron is described by the Dirac equation, which we will meet later;

---

[22] B. Odom et al 2006 Phys. Rev. Lett 97 030801

theoretical predictions based on the Dirac equation, give the value of the g-factor for the electron to be $g_s = 2$. Experiments measure the spin g-factor to be $g_s = 2.00231930419922 \pm (1.5 \times 10^{-12})$[23]. We understand why this is not exactly two; it is to do with virtual photons which we will consider later. For now, we continue taking the g-factor to be exactly two as predicted by the Dirac equation.

We see that an electron is twice as effective as a classical rotating charged body at producing magnetic dipole moment. However, the magnetic dipole moment of the electron is one Bohr magneton; experiments measure the electron to have one Bohr magneton of magnetic dipole moment. Thus, if we associate electron spin with classical angular momentum, then the electron has $\frac{\hbar}{2}$ angular momentum; that is $S = \frac{\hbar}{2}$. That is half of a quantum of classical angular momentum. Because of this, the electron is known as a spin one-half particle.

Aside: There are other spin one-half particles in physics, and there are spin one particles in physics which have a whole $\hbar$'s worth of angular momentum. The photon is a spin one particle. String theorists postulate a spin two graviton, but it has not been observed.

The intrinsic spin of an electron is thought of as being equivalent to classical angular momentum because, based on that equivalence, other than the factor of 2, the maths seems to work properly. However, since the electron is a point particle of zero radius, angular momentum in its classical form, $\vec{L} = \vec{r} \times \vec{p}$, makes no sense in the case of the electron. It seems that the electron's magnetic dipole moment is 'caused' by a non-classical type of angular momentum. It seems that we have magnetic dipole moment 'caused' by two different types of angular momentum; these are classical angular momentum and intrinsic spin angular momentum.

---

[23] Thank-you Wikipedia – Electron magnetic moment

How can we have two types of angular momentum? We could have two types of empty space each with its own type of rotation. We will introduce the reader to quaternion space in due course; the reader will meet 'double rotation' in quaternion space. Meanwhile we summarise the story so far:

i)   Electrons have one Bohr magneton of magnetic dipole moment $\mu_B = \dfrac{eh}{2m_e}$.

ii)   Electrons have an angular momentum, called intrinsic spin or just spin, of $\dfrac{\hbar}{2}$.

iii)   The electron g-factor is two.

iv)   Angular momentum is quantitised.

Above, we have outlined the magnetic properties and the quantitative spin properties of the electron. None of this was known in the early 20th century. It was in 1922, in Frankfurt, Germany, that Otto Stern (1888–1969) and Walther Gerlach (1889–1979) performed the experiment that eventually led to the discovery of the intrinsic spin of the electron.[24]

Before we look at the experiment done by Stern and Gerlach, remember that magnetic dipole moment has direction as well as amount. Above, we have not considered the direction of the electron's magnetic dipole moment. Similarly, angular momentum has direction as well as amount. The Stern-Gerlach experiment, as well as portending the discovery of intrinsic spin, produced an associated, but utterly unexpected, discovery.

The Stern-Gerlach experiment:
Stern and Gerlach sent neutral silver atoms through an inhomogeneous (not uniform) magnetic field in a well-defined direction. The silver atom is an electrically neutral particle, but,

---

[24] W Stern & O Gerlach. Z. Phys 8 110 (1921)

because the silver atom has a valence electron, it has a magnetic moment due to that electron; roughly, the non-valence electrons are paired together in a way such that their magnetic moments cancel; the single valence has no partner to cancel its magnetic moment. Because silver atoms have a magnetic moment, they will be deflected as they transit a non-uniform magnetic field. Since the silver atoms were produced with randomly oriented magnetic moments, it was expected that the deflections caused by the magnetic field would be in random amounts and random directions. Indeed, the Stern Gerlach experiment was done in the expectation that this would be the case for neutral atoms. This expected result is not what Stern and Gerlach found.

As the silver atoms transited the non-uniform magnetic field, they most surprisingly aligned either 'up' or 'down' with the z-axis of the co-ordinate system of the magnetic field. Half of the silver atoms aligned with the 'up' direction and half of the silver atoms aligned with the 'down' direction. This was not the random distribution that had been expected. Due to such definite 'only two directions' alignment, the silver atoms were deflected by the same amount either 'up' by the magnetic field or 'down' by the magnetic field, but there was no deflection in any other direction or by any other amount. The amounts of the deflections were the same because the amount of the silver atom's magnetic moment is a definite quantity. On a collecting screen, where Stern and Gerlach had expected to see a continuous smudge, they saw instead two distinct spots. The spots were a little blurred because the velocity of the silver atoms transiting the magnetic field varied a little[25], but there were two distinct spots. The deflection induced by the inhomogeneous magnetic field was discrete not continuous; this indicates that the magnetic dipole moment of the electron has (two) discreet directions.

This is not quantitisation of the amount of the silver atoms magnetic dipole moment. This is quantitisation of the spatial direction of the magnetic dipole moment of the silver atoms. It is one thing for each silver atom to have the same definite amount of magnetic dipole

---

[25] O. Stern Z. Phys. 2 49 (1920). & J. F. Zartman Phys. Rev. 37 383 (1931)

moment, but in discreet directions? How can that happen in our 4-dimensional space-time? In our space-time, we can point our fingers in all directions; we cannot imagine being able to point our fingers in only two diametrically opposed directions.

The experiment became known as the Stern-Gerlach experiment[26]. The apparatus, which is comprised of the source and the inhomogeneous magnetic field poles, has become known as the Stern-Gerlach apparatus or more often as a Stern-Gerlach machine.

We have a picture:

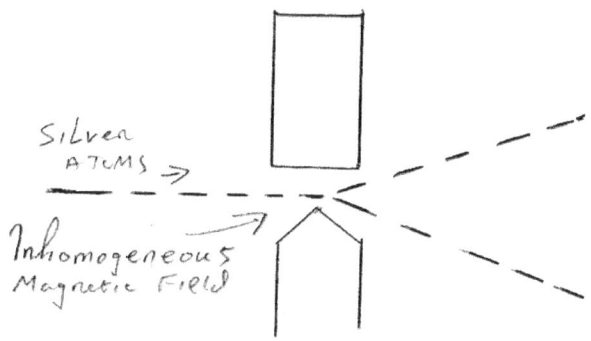

It is not practical to use electrons, or any other charged particle, in the Stern-Gerlach experiment due to the very large deflection caused by the electric charge, but we might think of the Stern-Gerlach experiment being done with electrons. In such a case, if we could account for the large deflection caused by the electric charge, we would discover that the magnetic moment of the electron is either 'up' or 'down' and never in-between. As far as the electron spin is concerned, spatial direction is either 'up' or 'down' but never in-between. This is an amazing thing to we who are accustomed to being able to point our finger in any direction. It seems that the electron is able to 'point its finger' in either only the 'up' direction or the 'down' direction[27].

---

[26] The Stern-Gerlach medal is the most prestigious German award for experimental physics.
[27] Electrons do not really have fingers even though, as we will see later, they are either left-handed or right-handed.

It gets more mysterious:

We have another picture[28]:

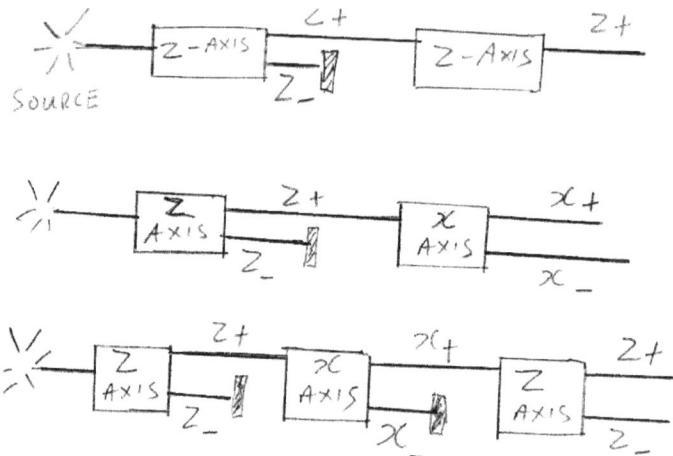

Looking at the picture above, we see that, if we have two Stern-Gerlach machines oriented in the same direction and we block the 'down' silver atoms, we get exactly what we would expect. Half the silver atoms emerge from the first Stern-Gerlach machine in the 'up' direction, and half the silver atoms emerge in the 'down' direction, but, because of the block on the 'down; silver atoms, only the 'up' silver atoms emerge from the second machine.

We might not be surprised to discover that when two orthogonally oriented Stern-Gerlach machines are in line we see the second machine separate the silver atoms into 'left' and 'right' in the x-direction.

When we put three Stern-Gerlach machines in line such that the first and the last machines are oriented in the same way but the middle machine is oriented orthogonally to the other two machines, we might think that, since we filtered out the z-direction 'down' silver atoms from the first Stern-Gerlach machine, we would get only z-direction 'up' silver atoms coming from the third Stern-Gerlach machine. How wrong we would be. In spite of ridding ourselves of the z-direction 'down' silver atoms from the first Stern-Gerlach

---

[28] Many thanks to Wikipedia.

machine, we see they are back as they emerge from the third Stern-Gerlach machine. The reader should take a moment to think about this. It is not what anyone would intuitively expect. What is happening here?

It gets even more curious; we have another picture:

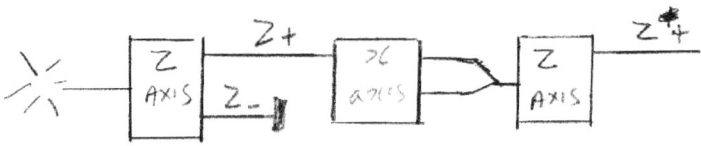

Suppose we filter out the z-direction 'down' silver atoms before we put them into the second Stern-Gerlach machine but we do not observe the output of that second machine and instead feed the whole of that output into another machine oriented as is the first machine. In this case, we have not observed the x-direction of the silver atoms. Remarkably, the silver atoms have 'remembered' their z-direction in spite of going through an x-direction Stern-Gerlach machine.

It is as if, when a silver atom is in a particular state, interaction in our 4-dimensional space-time 'knocks' the silver atoms out of that state, but that the silver atoms will remain in that state if not 'interfered with' by an interaction in our 4-dimensional space-time.

It is as if electrons 'sleep' in their own space until they are woken by interaction with our space-time. After the interaction, the electrons go back to 'sleep' in their own space but the bed is in a different direction from before they were 'woken' by the interaction with our space-time.

Probability again:
Although a Stern-Gerlach machine will divide a supply of silver atoms half into the 'up' states and half into the 'down' states, this is just a statistical average. For an individual silver atom entering a Stern-Gerlach machine, regardless of how the atom's magnetic dipole moment is oriented, the direction in which the atom will exit the Stern-Gerlach machine is dependent on chance. There is no

deterministic process by which the exiting silver atom can be predicted to be oriented in a particular direction. There is a calculable probability of a particular silver atom leaving a Stern-Gerlach machine oriented in a given direction. That probability depends upon the original orientation of the magnetic dipole moment of the silver atom. If the original magnetic dipole moment is in exactly the same direction as the Stern-Gerlach machine, then the probability of that atom exiting the Stern-Gerlach machine unchanged is 100%, as seen in the pictures above, but this is still a probability. Probability is in the physics here.

Chapter 6

Electron Spin Again

"It was a little over fifty years ago that George Uhlenbeck and I introduced the concept of spin ... most young physicists do not know that spin had to be introduced. They think it was revealed in Genesis or ... ", Samuel A Goudsmit – February 1976.

Electrons are particles of zero spatial size in our 4-dimensional space-time. It is thus meaningless to think of an electron rotating on an axis in our 4-dimensional space-time. Classically, angular momentum is $\vec{L} = \vec{r} \times \vec{p}$; since electrons have zero size, $r = 0$, and so the angular momentum of an isolated electron must be zero; except it isn't.

When Uhlenbeck (1900-1988) and Goudsmit (1902-1978) proposed that an isolated electron, as well as having a definite amount of mass and a definite amount of electric charge, also has a definite amount of angular momentum, they effectively proposed that the electron has a definite amount magnetic dipole moment.

But is there really angular momentum associated with the magnetic moment of our non-classical electron? Perhaps it is only magnetic moment. No! it is not only magnetic moment, there is angular momentum associated with our non-classical electron, and it corresponds to classical angular momentum, and this is clearly shown in the *Einstein de Haas* effect.

The Einstein de Haas effect:
Imagine a vertical iron rod hanging from a light line. The atoms in the rod are randomly aligned in such a way that the angular momenta of the differently orientated atoms cancel and the rod is still. Switch on a magnetic field around the rod, and the atoms in the rod 'snap' into alignment with that magnetic field, but this unbalances the

cancelation of the atomic angular momenta. Since angular momentum is conserved, the rod starts to rotate to compensate for the unbalancing of the atomic angular momenta; this is classical rotation associated with classical angular momentum. Atomic angular momentum has become classical angular momentum. This effect is very small, but it was demonstrated by Albert Einstein and W. J. de Haas (1878-1960) in 1915 and is called the Einstein de Haas effect.[29] Although the effect is very small, it has been measured accurately[30,31]

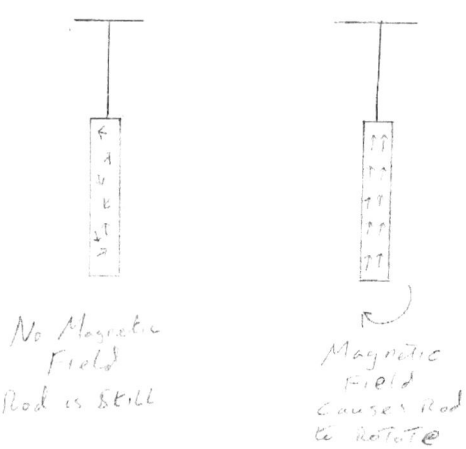

No Magnetic
Field

Rod is Still

Magnetic
Field
Causes Rod
to Rotate

The art just gets better.

The angular momentum of atomic systems is intimately entwined with the magnetic properties of those systems. The Einstein de Haas effect clearly links classical angular momentum to atomic magnetic moment. (It provides no evidence for angular momentum being quantitised.)

The photon is a spin one particle:
There is another consequence of the electron having angular momentum. Imagine an electron with its magnetic moment pointed 'down' being suddenly subject to a magnetic field that makes it

[29] The Einstein de Haas effect was predicted by O. W. Richardson in 1908 Phys. Rev. 26 248 (1908)
[30] J. Q. Stewart Phys. Rev. 11 100 (1918)
[31] A. P. Chattock L. F. Bates Phil..Trans. Roy. Soc. A 223 257 (1923)

'snap' into a magnetic moment 'up' position. There is a difference in energy between an electron aligned with a magnetic field and one that is aligned against a magnetic field; what happens to the energy that is the difference in energy between the two states? The difference in energy is taken away from, or brought to, the electron by a photon, but the 'snapping into position' changed the electron's angular momentum as well as its energy. If the electron's angular momentum was $-\dfrac{\hbar}{2}$ to begin with, it will be $+\dfrac{\hbar}{2}$ after it has flipped. As well as taking away the energy difference, the photon must also take away the difference in angular momentum. The difference in angular momentum between a 'down' electron and an 'up' electron is twice the difference in angular momentum between an electron and zero angular momentum. The photon therefore carries away twice the angular momentum of the electron; that is the photon has $\hbar$ amount of angular momentum. We say that an electron has spin one-half intrinsic angular momentum while the photon has spin one intrinsic angular momentum. We will see this presented as an electron is a spin half particle and a photon is a spin one particle. Since the photon has spin, the electron must have spin (conservation of angular momentum).

Neutrons and protons have magnetic moments:
Although the neutron is an electrically neutral particle, it has a magnetic moment due to the magnetic moments of the constituent quarks of which it is comprised. A neutron is two down quarks and one up quark. Similarly, and perhaps less surprisingly, the proton has a magnetic moment due to the magnetic moments of the constituent quarks of which it is comprised. A proton is two up quarks and one down quark. We have from detailed calculation of the wave functions of these particles:

$$\mu_{\text{Proton}} = \frac{1}{3}\left(4\mu_{up} - \mu_{down}\right) \qquad (6.1)$$

$$\mu_{\text{Neutron}} = \frac{1}{3}\left(4\mu_{down} - \mu_{up}\right) \tag{6.2}$$

$$\mu_{up} = -2\mu_{down} \tag{6.3}$$

This leads to:

$$\frac{\mu_{\text{Neutron}}}{\mu_{\text{Proton}}} = -\frac{2}{3} = -0.6666... \tag{6.4}$$

The experimentally measured value of the ratio of the magnetic moments of the proton and the neutron is[32]:

$$\frac{\mu_{\text{Neutron}}}{\mu_{\text{Proton}}} = -0.68497945 \pm 0.00000058 \tag{6.5}$$

Summary:

As well as a definite amount of electric charge and mass, the electron has a definite amount of intrinsic spin angular momentum and an associated definite amount of magnetic dipole moment.

i)   All electrons have half a quantum, $\frac{\hbar}{2}$, of angular momentum.

ii)   All electrons have one Bohr magneton of magnetic dipole moment.

iii)   Photons are spin one particles.

---

[32] Halzen & Martin: Quarks & Leptons: page 55

Chapter 7

A Note on Lepton Number

There are two different types of fundamental sub-atomic particles. The two types are known as bosons and fermions. Bosons are named after Satyendra Nath Bose (1894-1974) and fermions are named after Enrico Fermi (1901-1954).

Bosons:

Bosons are spin one particles like the photon. Bosons carry the electromagnetic force (photon), the weak nuclear force ($Z^0, W^{\pm}$), and the strong nuclear force (gluons). Within QFT, all forces are associated with a type of particle, boson, which carries the force, as momentum, between two particles that are subject to that force.

Within the conventional understanding[33], each boson corresponds to a generator in a unitary Lie group[34]. We list the correspondence for interest, but the reader should not be daunted if this list is meaningless to them at this time; we come back to this later in the book.

$$
\begin{array}{ll}
\textit{Photon} & U(1) \\
Z^0, W^{\pm} & SU(2) \\
\textit{8 gluons} & SU(3)
\end{array}
\qquad (7.1)
$$

$U(1)$ stands for Unitary Group of 1×1 matrices. $SU(2)$ stands for Special Unitary Group of 2×2 matrices. $SU(3)$ stands for Special Unitary Group of 3×3 matrices. The names are misleading in that it is only notation that determines the size of the matrices that represent these groups.

---

[33] Your author opines that $SU(3)$ does not really exist. It's only his opinion.
[34] See Dennis Morris: Lie Groups and Lie Algebras.

Each generator is a matrix such that the exponential of that matrix is a rotation matrix. Each generator generates a 2-dimensional Euclidean rotation matrix; for example, the $U(1)$ generator is:

$$\begin{bmatrix} 0 & 1 \\ -1 & 0 \end{bmatrix} \equiv [i] \qquad (7.2)$$

The exponential of this generator, (7.2), multiplied by a real number gives a 2-dimensional Euclidean rotation matrix:

$$\exp\left(\begin{bmatrix} 0 & \theta \\ -\theta & 0 \end{bmatrix}\right) = \begin{bmatrix} \cos\theta & \sin\theta \\ -\sin\theta & \cos\theta \end{bmatrix} \qquad (7.3)$$

There are three generators in $SU(2)$, and there are eight generators in $SU(3)$.

As we said earlier, string theorists postulate a spin 2 boson that carries the gravitational force, but there is no evidence of this postulated boson.

Bosons carry force. Fermions are matter.

Fermions:

Fermions are spin one-half particles like the electron. These particles are the constituents of matter. Single fermions are either leptons or quarks. There are two basic types of leptons, the electron with electric charge -1 and the neutrino with electric charge 0, but there are three generations of these two types; the other two generations are known as the muon and the muon neutrino and the tau and the tau neutrino. There are two types of quarks, the up quark and the down quark, with electric charge $\left(\dfrac{2}{3}, -\dfrac{1}{3}\right)$ respectively, but there are three generations of these two types; the other two generations are known as the charm and strange quarks and the bottom and top quarks. The complete list of fermions is:

$$e^- \quad v_e \quad : \quad \mu^- \quad v_\mu \quad : \quad \tau^- \quad v_\tau \qquad (7.4)$$
$$u \quad d \quad : \quad c \quad s \quad : \quad t \quad b$$

It is often said that the only difference between the generations of fermions, as far as we know, is the mass of the particles. That is not quite true. It is part of the Standard model of particle physics that different generations of fermions have different types of lepton number[35], and so generations of fermions differ from each other by both mass and by the nature of their lepton number. If, as it does later, it turns out that neutrinos are all of zero mass, then the three generations of neutrinos differ from each other in only the nature of their lepton number.

The bosons do not come in different generations; only one copy of each boson exists.

Anti-fermions:
Corresponding to each fermion, there is an anti-fermion. The anti-electron is called the positron denoted by e^+; the anti-neutrino is denoted by $\overline{v}_l$. In general, anti-particles are denoted by a bar above the usual sign. Similarly, the anti-quarks are denoted by a bar over the letter like $\overline{u}$. The bosons are said to be their own anti-particles, or perhaps they do not have anti-partners.

Within conventional quantum theory, fermions are mathematically represented by an ordered pair of complex numbers that is called a spinor.

Your author points out that a quaternion can be written as an ordered pair of complex numbers, and he opines that a fermion is best represented as a quaternion. We have:

---

[35] No-one has any understanding of the nature of lepton number.

$$e^- = \begin{bmatrix} a+ib \\ c+id \end{bmatrix} \equiv \begin{bmatrix} a & b & c & d \\ -b & a & -d & c \\ -c & d & a & -b \\ -d & -c & b & a \end{bmatrix} \tag{7.5}$$

The anti-fermion is just the conjugate of the fermion. We have used the electron and the positron to illustrate this:

$$e^+ = \begin{bmatrix} a-ib & -c+id \end{bmatrix} \equiv \begin{bmatrix} a & -b & -c & -d \\ b & a & d & -c \\ c & -d & a & b \\ d & c & -b & a \end{bmatrix} \tag{7.6}$$

The strange conjugation of the pair of complex numbers is an artefact of the notation; it is much simpler with quaternions.

Baryons and Hadrons:

Quarks will bind together (by the strong force) to form either pairs or triplets of quarks. The pairs of quarks are called mesons and have various names like kaons or pions. The triplets of quarks are called baryons and are particles like the proton, p^+, and the neutron, n. The mesons and baryons are collectively called hadrons because they 'feel' the strong force – phew!

The Standard model:

Our understanding of the particles of our universe and how they interact is a theory known as the 'Standard model'. It is often said that the Standard model is enormously successful in that it accurately predicts all particle physics phenomena except the masses of neutrinos. That is not quite true; for example, the Standard model does not predict the directional quantitisation of electron spin, the value of physical constants, or that the universe is 4-dimensional. I am sure we could find other phenomena it does not predict if we looked hard enough. It is accepted and widely proclaimed that the Standard model does not predict the particle content of the universe.

Why do we have the numbers and types of particles that we observe within our universe? We have not a clue. The particle content of the Standard model is assumed.

A little history:

In your author's personal view, it is a little unnerving to realise how many years have passed since sub-atomic particles were first discovered – how old is this science! In your author's view, it is a little unnerving to realise how recently the sub-atomic particles were discovered – how new is this science!

Particle	Theoretical date of discovery	Experimental date of discovery
Electron	-	1897
Positron	1928	1931
Proton	-	1911
Anti-proton	1928	1955
Neutron	1920	1932
Anti-neutron	1928	1956
Neutrino	1930	1954
Anti-neutrino	1930	1954

We have known about electrons for over a hundred years, and still we do not understand them. Your author is old enough to have been born when anti-protons had still not been experimentally discovered.

Lepton number:

In the early 1950's, physicists were puzzled by the absence of nuclear reactions such as:

$$\bar{v} + n \rightarrow p^+ + e^- \qquad (7.7)$$

43

This symbolises a neutron, n, and an anti-neutrino, $\bar{\nu}$, charging into a proton, p^+, and an electron, e^-. Physicists saw nuclear reactions like:

$$\bar{\nu} + p^+ \rightarrow n + e^+ \tag{7.8}$$

Why one reaction but not the other? In 1953, E. Konopinski (1911-1990) and H. Mahmoud introduced the concept of conservation of lepton number[36] to explain these observations.

Leptons are assumed to have a lepton number of plus 1 while anti-leptons are assumed to have a lepton number of minus 1 and non-leptons are assumed to have a lepton number of zero. So, an electron has a lepton number of +1; a positron (anti-electron) has a lepton number of −1; a neutrino has lepton number +1; an anti-neutrino has lepton number −1. With these assignments, conservation of lepton number explains the observed reactions and absence of reactions (7.7) & (7.8) above.

It seems as if lepton number is some kind of charge like electric charge, but we do not understand it properly.

Note: It used to be universally thought that the neutrinos were massless. This implied that lepton number was conserved within each separate lepton generation, and so we would have three separate lepton number conservation laws. It is now thought, though not universally accepted, that neutrinos have some kind of mass (there might be different kinds of mass) and this implies that lepton number is conserved over all three generations of leptons but not within a single generation. The Standard model assumes that lepton number is conserved within each separate lepton generation. It is by this assumption that the Standard model assumes that neutrinos are massless. We will look at this in more detail later.

---

[36] E. Konopinski & H Mahmoud "The Universal Fermi Interaction" Phys. Rev. 92 (1953) 1045

Summary:

We now have to add to our list of electron properties that electrons have lepton number +1. Of course, electron neutrinos also have lepton number +1.

More than that, we need to distinguish between the three different types of lepton number corresponding to the three generations of fermions. We need to amend the above paragraph.

We now have to add to our list of electron properties that electrons have electron generation lepton number +1. Of course, electron neutrinos also have electron generation lepton number +1.

Chapter 8

Quaternion Space – A Road Less Travelled

"... Two roads diverged in a wood, and I took the one less travelled..." – Robert Frost

Throughout history, humankind have believed we understood the universe. Our most ancient understandings involved great white spirits or other gods and goddesses atop mountains or in woodlands. More recent understandings include belief in the luminiferous ether or the belief in the separate nature of space and time. When we look back with hindsight, we are amazed and bewildered that intelligent individuals, even physicists and mathematicians, believed such theories in the depth of their hearts. Edward Morley[37] (1838-1923) really did believe in the luminiferous ether, and, it seems, Isaac Newton really did believe in God.

Today, we have concepts that invoke questions that we hardly dare ask because such questions deeply unsettle the confident world understanding that is based on those concepts. For example, quantum physicists tell us that complex Hilbert spaces are central to our universe. These are spaces with complex axes rather than real axes. Will we look back fifty years from now with the view that this is obvious nonsense and be in amazement that anybody swallowed such nonsense? It is the nature of humanity that we dream we are awake.

In this chapter, we embark upon a change of understanding; we take *"...the road less travelled...",* and we begin a new dream.

In this chapter, we introduce the reader to quaternion space. We will contrast this with our familiar 4-dimensional space-time. Both quaternion space and our familiar 4-dimensional space-time derive from the $C_2 \times C_2$ finite group, and so they are closely connected. Our

---

[37] Famous for the Michelson Morley experiment of 1887.

4-dimensional space-time derives from the six A_3 algebras of the $C_2 \times C_2$ group[38]. Quaternion space derives from the two quaternion algebras of the $C_2 \times C_2$ group. There are no other non-commutative algebras within the commutative $C_2 \times C_2$ group.

We will assert that the electron exists in quaternion space and that this explains the 'weird' nature of the electron. In later chapters, we will further consider our assertion as we look at other aspects of the electron. To be honest, we are beginning to wonder if the electron actually is quaternion space.

The reader is warned that it is not standard mantra to view the electron as existing within quaternion space. We are about to leave the path of accepted understanding and strike out in a new direction. The reader might see this as no more than an interesting stroll through unexplored countryside, or the reader might feel there is something of substance here. Of course, if the reader is entirely happy with the ephemeral nature of electrons of no size with magnetic dipole moment of quantitised direction popping in and out of our space-time whenever they feel like it, then the reader might not wish to leave the well-walked path.

Our 4-dimensional space-time:
Often in science, progress is made by asking a question that before no one thought to ask. Let us try doing that.

Why can the reader wave her arms around?

The reader can wave her arms around because it is possible to rotate in all six 2-dimensional planes within our 4-dimensional space-time. Three of these 2-dimensional planes are 2-dimensional Euclidean planes[39], and three of these 2-dimensional planes are 2-dimensional

---

[38] See: Dennis Morris: Upon General Relativity
[39] All Euclidean planes are 2-dimensional, sorry.

space-time planes[40] in which rotation is just a change of velocity often known as a Lorentz boost.

A 4-dimensional space has four axial variables (axes), and there are six ways to choose two variables from four. These six pairings of axial variables correspond to six 2-dimensional planes. We have six angles from the six A_3 algebras that together form our 4-dimensional space-time; that is one angle for each 2-dimensional plane, and so we can rotate in all six 2-dimensional planes.

A space in which we can rotate in all possible 2-dimensional planes is called a geometric space[41]. There is only one geometric space as observation will confirm and calculation will predict. Our 4-dimensional space-time is that unique geometric space[42].

In short, the reader can wave her arms around because she can rotate in the six 2-dimensional planes of our 4-dimensional space-time.

A pretend space:
Now, let us consider the familiar 3-dimensional Euclidean space that is a sub-space of our 4-dimensional space-time. Imagine the reader could rotate in only two of the three Euclidean 2-dimensional planes of our 4-dimensional space-time. Imagine the reader could rotate in both the orthogonal vertical 2-dimensional planes but not in the horizontal plane. Imagine that this is because there are only two angles in the space and that these angles apply to the two vertical rotational planes.

---

[40] All space-time planes are also 2-dimensional, but we distinguish between 4-dimensional space-time and 2-dimensional space-time.

[41] Within the last year, your author's use of the term 'geometric space' has changed. Within the book 'The Physics of Empty Space' your author uses the term 'geometric space' where he ought to have used the term 'spinor space'. Your author apologises for any confusion caused.

[42] See: Dennis Morris: The Uniqueness of our Space-time.

The possible rotations available are two vertical circles (hoops) at right angles to each other intersecting at only two points (top and bottom).

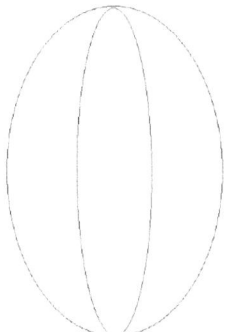

The art just gets better and better.

Now, suppose the reader tries to rotate in both of these planes at the same time. Such a joint rotation necessarily includes a bit of rotation in the horizontal plane; indeed, after 90^0 worth of such joint rotation, one would find oneself at 45^0 to each vertical hoop as if one had effectively rotated in the horizontal plane by 45^0. Let us just say that again; a joint rotation necessarily includes a bit of rotation in the horizontal plane – this bit is important. This geometric fact was first proved by the Swiss mathematician Leonard Euler (1707-1783). In Euler's proof, this geometric fact takes the form: "Any combination of two (2-dimensional) rotations in $\mathbb{R}^3$ is a single (2-dimensional) rotation."

The difficulty is, by definition of our pretend space, that we cannot rotate in the horizontal plane; we have no horizontal angle to measure rotation in the horizontal plane. If we cannot rotate in the horizontal plane, we cannot compose together rotations in the two other planes into a rotation.

The point is that, when there are three possible 2-dimensional rotational planes but there are only two angles, it is possible to rotate in only two rotational planes. In such a space, we cannot compose together rotations in those two planes. We cannot wave our arms around freely in such a space. Yes, I know we are working in a pretend space, but we will shortly see that something very similar to this is the case in quaternion space, and quaternion space is not a pretend space. Back to the pretend space.

The angular momentum vector in such a pretend space would point in one of four possible directions only; these being perpendicularly left or right with respect to each of the two rotation planes. The angular momentum vector could not point in all possible directions; it could point only either 'up' or 'down' with respect to the rotational plane – just like the spin of an electron.

Note that changing the co-ordinate system would not change this basic geometric fact. Basic geometric facts are independent of the arbitrary co-ordinate system chosen by we mere mortals.

You cannot wave your arms in quaternion space:
We have the quaternion rotation matrix:

$$
\begin{bmatrix}
\cos\lambda & \dfrac{\lambda}{b}\sin\lambda & \dfrac{\lambda}{c}\sin\lambda & \dfrac{\lambda}{d}\sin\lambda \\[2ex]
-\dfrac{\lambda}{b}\sin\lambda & \cos\lambda & -\dfrac{\lambda}{d}\sin\lambda & \dfrac{\lambda}{c}\sin\lambda \\[2ex]
-\dfrac{\lambda}{c}\sin\lambda & \dfrac{\lambda}{d}\sin\lambda & \cos\lambda & -\dfrac{\lambda}{b}\sin\lambda \\[2ex]
-\dfrac{\lambda}{d}\sin\lambda & -\dfrac{\lambda}{c}\sin\lambda & \dfrac{\lambda}{b}\sin\lambda & \cos\lambda
\end{bmatrix}
\tag{8.1}
$$

$$\lambda = \sqrt[2]{b^2 + c^2 + d^2}$$

This is rotation in a 4-dimensional space, but there are only three angle variables, $\{b,c,d\}$. Because this is a 4-dimensional space, there are six ways to pair two axes together. Because there are only three angle variables, $\{b,c,d\}$, we can rotate in only three 2-dimensional planes in quaternion space. That is worth repeating; we can rotate in only three 2-dimensional planes in this 4-dimensional quaternion space.

There are only two possible types of rotation in a 2-dimensional plane; these are 2-dimensional Euclidean rotation and 2-dimensional space-time rotation (Lorentz boost). We take only one variable by setting the other two variables to zero. This is no more than a change of co-ordinate system. This gives:

$$\begin{bmatrix} \cos\lambda & \dfrac{\lambda}{b}\sin\lambda & 0 & 0 \\[2ex] -\dfrac{\lambda}{b}\sin\lambda & \cos\lambda & 0 & 0 \\[2ex] 0 & 0 & \cos\lambda & -\dfrac{\lambda}{b}\sin\lambda \\[2ex] 0 & 0 & \dfrac{\lambda}{b}\sin\lambda & \cos\lambda \end{bmatrix} \qquad (8.2)$$

$$\lambda = \sqrt{b^2}$$

Cancelling the λ in (8.2), we see that the nature of this rotation is Euclidean - look at the type of trigonometric functions; they are Euclidean trigonometric functions and they are not hyperbolic trigonometric functions. The other two variables also give a Euclidean type of rotation.

Putting it in a more technical way, the 4-dimensional quaternion algebra has only three 2-dimensional sub-algebras and each of these 2-dimensional sub-algebras is algebraically isomorphic to the Euclidean complex numbers, $\mathbb{C}$. Ultimately, quaternions are like this because the group $C_2 \times C_2$ from which the quaternions derive has three C_2 sub-groups. The three sub-groups correspond to the three sub-algebras that correspond to the three rotation planes.

Note: Algebraically isomorphic does not mean identical; algebraically isomorphic means identical algebraic properties; there might be geometric properties or other properties that differ.

Because there are only three angle variables in quaternion space but there are six possible 2-dimensional planes in every 4-dimensional space, the rotational surface in this 4-dimensional quaternion space is not a spherical surface but is three intersecting circles. We can rotate only in the three planes. We cannot wave our arms about in this space. We have a match for the quantitised spatial direction nature of electron spin.

Let us take a breather. When, in 1922, it was first discovered that electrons have angular momentum which is directionally quantitised into pointing either 'up' or 'down' in three different directions, physicists were, to use a colloquialism, gobsmacked. It seemed impossible to imagine any mechanism which could explain this directional quantitisation within our 4-dimensional space-time. 'Seemed' is not expressive enough; it is impossible to have directional quantitisation in our 4-dimensional space-time – just look around. We have seen above that, within quaternion space, directional quatitisation is entirely natural. You see why we set out upon this '...*road less travelled*...'.

Of course, if we accept, as everyone does, that directional quantitisation does not happen in our 4-dimensional space-time, then we must accept that the electron is not in our 4-dimensional space-time.[43]

Double cover:

It has been known for many years that the magnetic moment of the electron is twice what we would expect for the same amount of classical angular momentum; the electron spin g-factor is 2, to be technical.

Looking at the quaternion rotation matrix above, (8.2), we see there are two rotations, one clock-wise and one anti-clockwise within this quaternion rotation – see the position of the minus signs. We offer another single variable:

---

[43] Did you see how surreptitiously your author slipped that profound realisation into the text?

$$
\begin{bmatrix}
\cos\lambda & 0 & \dfrac{\lambda}{c}\sin\lambda & 0 \\[2ex]
0 & \cos\lambda & 0 & \dfrac{\lambda}{c}\sin\lambda \\[2ex]
-\dfrac{\lambda}{c}\sin\lambda & 0 & \cos\lambda & 0 \\[2ex]
0 & -\dfrac{\lambda}{c}\sin\lambda & 0 & \cos\lambda
\end{bmatrix}
\qquad (8.3)
$$

$$
\lambda = \sqrt[2]{c^2}
$$

Cancelling the λ leads to a rotation twice in the clock-wise direction[44]. Presumably, twice the rotation warrants twice the magnetic dipole moment.

Within our 4-dimensional space-time, we see only single cover 2-dimensional rotations. How might a double cover quaternion rotation appear to us? Seemingly, we see twice the rotation we might expect.

The reader might often hear physicists speaking of the electron requiring $4\pi's$ worth of rotation to return to from where it started. This might be phrased like, "We must rotate the electron through 720^0 to rotate completely all the way back to where we started.". Well, looking at (8.2), we see we need 360^0 for the clockwise rotation and 360^0 for the anti-clockwise rotation to get back to where we started.

Does this double rotation within quaternion space explain the spin g-factor of the electron? The reader will form their own opinion.

More upon quaternion rotation:
Consider a rotation matrix within the 3-dimensional Euclidean part of our 4-dimensional space-time; we have:

---

[44] The direction might be anti-clockwise rather than clockwise, but who cares?

$$\begin{bmatrix} \cos\theta & \sin\theta & 0 \\ -\sin\theta & \cos\theta & 0 \\ 0 & 0 & 1 \end{bmatrix} \qquad (8.4)$$

This is rotation about an axis. Mathematically, we can show this by taking the eigenvectors of this matrix, (8.4). These are:

$$\begin{bmatrix} e^{i\theta} \\ 0 \\ 0 \end{bmatrix}, \quad \begin{bmatrix} 0 \\ e^{-i\theta} \\ 0 \end{bmatrix}, \quad \begin{bmatrix} 0 \\ 0 \\ 1 \end{bmatrix} \qquad (8.5)$$

We see that we get one eigenvector which is independent of the variable θ - it does not change as θ changes, and so this is a fixed axis of the rotation. We all know that every 2-dimensional rotation is about an axis; we are all wrong. Rotation in the 2-dimensional complex plane is not about an axis – there is no third dimension in the complex plane to be an axis. 2-dimensional rotation in our 4-dimensional space-time is rotation about two axes – two axes are independent of the angle variable. We have one axis of rotation above, (8.5), because we chose a 3-dimensional sub-space of our 4-dimensional space-time.

If we take the eigenvectors of the quaternion rotation matrix, (8.1), we find there are no eigenvectors which are independent of the angle variables. The same is true even if we reduce the quaternion rotation to a single variable. Using the quaternion rotation matrix, (8.2) or (8.3), we find there are no eigenvectors which are independent of the angle variable. Quaternion rotation is not rotation about an axis. In general spinor rotations are not rotations about an axis. That seems weird to us, but the maths is simple and straightforward; there are no eigenvectors which are independent of the angle variables. This needs to be repeated.

Look, the complex plane, $\mathbb{C}$, is 2-dimensional; there is no third dimension sticking out of the complex plane, and so rotation in the complex plane is 2-dimensional rotation in 2-dimensional space and there is no spare axis to be an axis of rotation. Our 4-dimensional space-time has 2-dimensional rotations, and so there are two spare dimensions which are unaffected by the 2-dimensional rotation in our 4-dimensional space-time. Quaternion space is a spinor space, just as the complex plane, $\mathbb{C}$, is a spinor space, and spinor spaces do not have rotation about an axis – this is the nature of rotation in a spinor space. It seems that spinor rotation is rotation about a point, the origin, rather than about an axis.

Now, the classical definition of angular momentum is $\vec{L} = \vec{r} \times \vec{p}$ wherein $\vec{r}$ is the distance from the axis of rotation. If we have no axis of rotation, as is the case in quaternion space, this definition seems meaningless. Well, perhaps the definition is correct in quaternion space but the r variable is the distance from the point of rotation rather than from an axis of rotation.

It would seem that intrinsic spin is rotation about a point rather than about an axis and that we have $\vec{L} = \vec{r} \times \vec{p}$ except that the r variable is the distance from the point of rotation rather than from an axis of rotation. The reader will form their own opinion.

Clearly, angular momentum in quaternion space is different from angular momentum in our 4-dimensional space-time, but then, the intrinsic spin of an electron is different from classical angular momentum.

Aha! Intrinsic spin is rotation that is not rotation about an axis. It is spinor rotation about a point. I knew there was something about intrinsic spin that made it different from what we call classical angular momentum.

Quantitised amounts of angular momentum:
The distance function within quaternion space is:

$$dist^2{}_{\mathbb{H}} = t^2 + x^2 + y^2 + z^2 \qquad (8.6)$$

There are no minus signs in the quaternion expectation distance function,(8.6). This is a space in which all axes are spatial; there is no time in this space. Now, angular momentum is associated with rotation. Without time, what is the difference between a slowly rotating object and a rapidly rotating object? – None. If we accept that an electron has an amount of angular momentum, intrinsic spin, within quaternion space, then within quaternion space, that amount of angular momentum will always be the same while-ever the mass is the same because the concept of slow rotation or fast rotation is meaningless in a timeless space.

Does this explain why the amount of intrinsic spin is quantitised? Seemingly, to your author, not very well. The reader will form their own opinion.

Quaternion rotation is 4-dimensional rotation:
The trigonometric functions of a 2-dimensional spinor space like the complex plane, $\mathbb{C}$, are such that they accept only one angle variable. We have:

$$\begin{bmatrix} \cos\theta & \sin\theta \\ -\sin\theta & \cos\theta \end{bmatrix} \tag{8.7}$$

The trigonometric functions of the 3-dimensional spinor spaces accept two angle variables[45]. The trigonometric funtions of 4-dimensional spinor spaces accept three angle variables – see (8.1). It is the nature of rotation in spinor space that it is N dimensional where N is the dimension of the spinor space. Of course, it must be like this if the rotation is not about an axis.

Rotation in quaternion space has a 4-dimensional nature; the quaternion trigonometric functions accept three angle variables. Even when we reduce the quaternion rotation to a single variable, we do not destroy the 4-dimensional nature of quaternion rotation; we merely re-align the co-ordinate system. Of course, we mortals who observe in our 4-dimensional space-time cannot see this 4-dimensional rotation. All we can see are the 2-dimensional rotations

---

[45] See: Dennis Morris Complex Numbers The Higher Dimensional Forms.

that are within our 4-dimensional space-time. We are unsure how quaternion rotation would appear in our 4-dimensional space-time, but it seems to manifest itself as double cover 2-dimensional rotation, just like intrinsic spin.

Point particle:

In the next few paragraphs, we are going to put to the reader an idea that we have found beside *"... the road less travelled ..."*.

Clearly, your author would not be presenting this idea if he did not think this idea has some merit. None-the-less, the reader is urged to be cautious and sceptical of this idea. This is certainly not the standard mantra taught in universities around the world. We begin to present our *"... road less travelled ..."* idea.

There are two quaternion algebras, which we call the quaternions and the anti-quaternions[46]. Since both are quaternion algebras, they have the same norm. This is established fact.

As above, the distance function, norm, within quaternion space is:

$$dist^2 = t^2 + x^2 + y^2 + z^2 \tag{8.8}$$

The emergent expectation distance function of the quaternion algebras is the sum of the distance functions of the two quaternion algebras and is of the same form as (8.8):

$$dist^2 = t^2 + x^2 + y^2 + z^2 \tag{8.9}$$

Compare this, (8.9), to the emergent expectation distance function of the A_3 algebras which is our 4-dimensional space-time[47]:

$$dist^2 = t^2 - x^2 - y^2 - z^2 \tag{8.10}$$

Now comes *"... the road less travelled ..."* idea.

---

[46] The left-chiral quaternions and the right-chiral quaternions is more appropriate.
[47] See Dennis Morris Upon General Relativity

57

Imagine we sit in not only one type of space but that we sit in two types of space. The two types are the emergent expectation quaternion space with distance function (8.9) and the emergent expectation A_3 space which is our 4-dimensional space-time and has distance function (8.10). We will take it that a single point, say, (a,b,c,d), exists in both spaces. Now let us take two such 'two-spaces' points and calculate the distance between them using the distance functions of the two emergent expectation spaces. We will get two numbers for the distance corresponding to the distance through the quaternion expectation space and the distance through the 4-dimensional space-time that is the A_3 expectation space.

The distance between two points in these two spaces will coincide only if the two distance functions give the same number when the variables are fed into them. This can happen in two possible instances.

The first of these instances is if we have the spatial variables $\{b,c,d\}$ equal to zero; in this case the two distance functions reduce to simply $dist^2 = a^2$. In our 4-dimensional space-time, we equate the a variable with time. The spatial extent is zero. This is a point in our 4-dimensional space-time with zero size, but it has duration.

The second of these instances is if we have the variable $\{a\}$ equal to zero; in this case, ignoring the signs, the two distance functions reduce to simply $dist^2 = b^2 + c^2 + d^2$. In our 4-dimensional space-time, this is spatial extent, but the temporal extent is zero. We have spatial extent that exists for zero time.

We see that, if the electron is a bit of quaternion space, then, because the distance functions differ as they do, the spatial size of that bit of quaternion space observed by an observer to endure within our 4-dimensional space-time will be of zero size – a point particle. Further, if that bit of quaternion space endures for zero time, then it will have spatial extent. The bits of quaternion space will appear in our 4-dimensional space-time to be either point particles which endure or to be spatially extensive objects which exist only instantaneously, while passing through two spatially separated slits in a screen perhaps.

In earlier chapters, we have pointed out that the electron is both:

1) a point particle which endures and is said to have particle-like properties
2) a spatially extensive object when it passes through two holes in a screen, interferes with itself, and is said to have wave-like properties.

This 'two spaces together' idea seems to give us something like electrons.

This is the end of the presentation of the *"... road less travelled ..."* idea. Be wary of new ideas.

Probalbility:
In our consideration of Stern Gerlach machines above, we met statistical probability associated with the spin direction of each emerging electron, and we asked what probability is doing in physics. We met probability also when we sent a single electron through a two holed screen in a cathode ray tube and we asked at what point would the electron hit the fluorescent screen.

For the last century, philosophers and physicists have pulled out their hair as they struggled to understand what probability is doing in physics. Albert Einstein famously rejected probability with his oft quoted words, "God does not play dice". After much brain wrenching tussle, it does seem that we have to accept that probability plays a part in our physical universe, yet still we are unsettled by this realisation.

Back to the *"... road less travelled ..."* idea. Let us consider the two separate emergent expectation spaces quaternion space and our 4-dimensional space-time. They each have their own co-ordinate systems, but the two co-ordinate systems are independent of each other. There is no connection or causal influence between the two separate spaces because they are separate spaces – separate universes, if you like. We can assert that the co-ordinate axes of our 4-dimensional space-time remain fixed with respect to ourselves in our laboratory, but we have no control over the co-ordinate axes of quaternion space which we must presume drift around at random with respect to our space-time co-ordinate axes.

Now imagine this quaternion space coincides with our space-time from point to point either as a point particle with duration or as a spatially extensive object with zero duration; this is what we described above. There is no way for an observer in our space-time to be able to predict what the orientation of the quaternion co-ordinate system will be at these coincidences. There is no causal connection between the separate spaces, and so we have lost determinism within our universe.

If it be that there is some order in the quaternion space[48], then there will be statisical order to the observations of quaternion space made by an observer in our space-time. This is exactly what physicists observe in our 4-dimensional space-time.

In which direction will an electron in quaternion space emerge from a Stern Gerlach machine? It is all a matter of probability because the co-ordinate system of quaternion space is free to drift randomly with respect to the co-ordinate system of our 4-dimensional space-time.

Would the above convince Einstein to accept probability in our universe? There are enough experimental results to prove we do have probability in our universe, but is the above a reasonable explanation of why we have probability in our universe? If we have got this correct, or even if we have pointed the direction for some other to understand why we have probability in our universe, then we have solved one of the greatest conundrums of 20th century physics. The reader will form their own opinion.

---

[48] Since the quaternions are a division algebra, there is much order in quaternion space.

Identical electrons:

Quaternion space is a mathematical structure. Every copy of it is identical. We will probably never know whether there exists only one copy of quaternion space which touches our 4-dimensional space-time at every interaction involving an electron or whether there are an infinite number of copies of quaternion space. Which-ever is correct, if electrons and quaternion space are the same thing, then all electrons will be identical. This is what we observe, of course.

Electron waves:

We point out that the quaternion trigonmetric functions that we see within the quaternion rotation matrix, (8.1), are wave-like functions. How would we in our 4-dimensional space-time which holds only 2-dimensional rotations with the associated 2-dimensional trigonometric functions see a quaternion trigonometric function?

Conservation laws:

The dynamics of our 4-dimensionsional space-time are constrained by the conservation of energy law and by the conservation of momentum law. The conservation of energy and the conservation of momentum are built upon the concept of time. These conservation laws say that the amount of energy and the amount of momentum do not change through time.

There are no minus signs in the distance function of quaternion space and so there is no time in quaternion space. If there is no time in quaternion space, it is meaningless to say that energy or momentum are conserved over time within quaternion space.

Clearly, each type of space has its own dynamics. Hm! I'm not sure that is clear at all, but let us assume it is true. We might expect some form of violation of conservation of energy and violation of conservation of momentum to be associated with an interaction between quaternion space and our 4-dimensional space-time.

Quantum field theory, QFT, is founded on such violations in the form of virtual particles with which we will deal later.

Right-chiral quaternions and left-chiral quaternions:

We mentioned above that there are two quaternion algebras which derive from the $C_2 \times C_2$ group. We will look at the differences between these two quaternion algebras later. This chapter has been a long chapter, and we do not wish to tire the gentle reader. Here, we merely make mention of the fact that quaternions come in both left-chiral (roughly left-handed) and right chiral (roughly right-handed) forms.

Summary:

In this chapter, we have tried to explain the nature of the electron by assuming the electron exists in quaternion space or perhaps is a bit of quaternion space. Have we succeeded?

We have had very good success with the quantitisation of electron spin direction. We also have a plausible explanation of the value of the spin g-factor, and of the difference between classical angular momentum and the intrinsic spin of the electron. Our explanation of the infinitesimally small point nature of the electron seems to work, but perhaps the explanation of why we have probability in our universe is unconvincing.

It is always the case that human understanding does not spring fully formed from the womb of a new world viewpoint. Even today, we are still refining Einstein's general relativity and Dirac's quantum field theory. Human understanding changes from decade to decade as new evidence is discovered or as old evidence is reinterpreted. It is unlikely that we will get everything right at our first attempt, and so the reader is warned to be cautious in accepting everything we have outlined above.

None-the-less, this is the first ever presented explanation of why electron spin is directionally quantitised.

Chapter 9

Electric Charge and Photons

Quantum field theory, QFT, sees electrons as excitations of the electron field rather than as particles or waves.[49]

Imagine a flat rubber sheet stretched horizontally. Now imagine a pair of fingers of someone's hand pinching the rubber sheet at a point and lifting it so that a little pimple is made in the rubber sheet. When the pinch is released, the pimple in the sheet vibrates up and down sending out waves like the waves on the surface of a pond consequent to a pebble being thrown into the pond. Of course, in our classical universe, the virations die away as friction and other extraneous forces deplete the energy of the waves, but, if there were no extraneous forces, the pimple would vibrate forever. We might say that the pimple is an excitation of the rubber sheet. QFT sees the rubber sheet to be equivalent to an electron field and sees the electron as an excitation, pimple, of that electron field.

This concept brings with it the idea that the electron field stretches throughout the entire 4-dimensional space-time universe and that each electron is a separate excitation of that electron field.

Similarly, photons are seen as excitations of the photon field, protons as excitations of the proton field, which is really three quark fields, and all other particles are similarly excitations of a field. In QFT, there are as many types of particle-field as there are types of particle, and every particle is the excitation of the appropriate particle-field. Since most particles are formed of quarks, at a fundamental level, QFT sees only lepton and quark fields and the associated boson fields.

---

[49] We wonder if the electron field is just the quaternion emergent expectation space.

What is a field?

There are different types of field.

A scalar field is a function of the space-time co-ordinates which associates a number with each point in 4-dimensional space-time; the temperature distribution of air within a room is an example of a scalar field.

A 4-dimensional vector field is a set of four functions of the space-time co-ordinates. The four functions together define a direction and an amount, a vector, which varies from point to point in space-time. A magnetic field is an example of a vector field.

A tensor field is a set of vectors, a tensor, defined at each point in space-time. In general relativity, the gravitational field is a tensor field.

In the conventional view, there are also spinor fields. In the conventional view, the electron field is a spinor field.

A spinor field is really a division algebra, like the quaternions or the complex numbers, whose components vary from point to point in space-time. For example, a quaternion field, that's a type of spinor field, over 4-dimensional space-time is of the form:

$$\begin{bmatrix} \phi(t,x,y,z) & A_x(t,x,y,z) & A_y(t,x,y,z) & A_z(t,x,y,z) \\ -A_x(t,x,y,z) & \phi(t,x,y,z) & -A_z(t,x,y,z) & A_y(t,x,y,z) \\ -A_y(t,x,y,z) & A_z(t,x,y,z) & \phi(t,x,y,z) & -A_x(t,x,y,z) \\ -A_z(t,x,y,z) & -A_y(t,x,y,z) & A_x(t,x,y,z) & \phi(t,x,y,z) \end{bmatrix}$$

$$(9.1)$$

In conventional QFT, the electron field is a pair of complex number fields. In conventional QFT, a pair of complex numbers is seen as a spinor field:

$$\begin{bmatrix} a(t,x,y,z)+ib(t,x,y,z) \\ c(t,x,y,z)+id(t,x,y,z) \end{bmatrix} \qquad (9.2)$$

This is deceptive, and, in your author's opinion, poor notation. It is a mathematical fact that a quaternion can be written as a pair of complex numbers; we have to distort the algebraic operations a little, but it can be done. In this book, the '... *road less travelled* ...' we will be taking assumes the electron field is a quanternion field over space-time as presented above, (9.1). This is not the conventionally assumed form of the electron field; the conventionally assumed form of the electron field is (9.2).

The Dirac equation – introduction:

For the purposes of this chapter, we will take an electron field to be the pair of complex number fields, (9.2). This is exactly what the Dirac equation takes the electron field to be.

Within quantum field theory, the equation which describes the electron, and all fermions, is the Dirac equation. The Dirac equation is derived from the Dirac lagrangian. A lagrangian is a mathematical expression. We will look at the Dirac equation in different notation and in more detail later. For now, we will use the standard notation, and we need to know only that the Dirac lagrangian is:

$$\mathcal{L}_{Dirac} = i\bar{\psi}\gamma_\mu\partial^\mu\psi - m\bar{\psi}\psi \tag{9.3}$$

The Dirac equation describes both the electron field and the positron field. Within the Dirac equation, the electron field and the positron field are represented by two pairs of complex numbers:

$$\psi = \begin{bmatrix} \begin{bmatrix} a+ib \\ c+id \end{bmatrix} \\ \begin{bmatrix} e+if \\ g+ih \end{bmatrix} \end{bmatrix} \equiv \begin{bmatrix} e^- \\ e^+ \end{bmatrix} \tag{9.4}$$

The symbol $\bar{\psi}$ is used to denote the adjoint electron and positron field. The complex fields are dependent upon the position in space-time $\psi(x) \equiv \psi(t,x,y,z)$. All eight independent variables,

$\{a,b,c,d,e,f,g,h\}$, are functions of the 4-dimensional space-time position, (t,x,y,z).

It is normal and accepted notation to put a single variable, x, within the parenthesis to indicate the four space-time variables (t,x,y,z), and so we write $\psi(x)$ to mean $\psi(t,x,y,z)$.

Symmetries and conservation laws:

The physics of the universe is the same in London as is the physics of the universe in Whitby. Change of position in space, spatial translation to be technical, does not change the physics of a system. A change of some form which leaves the physics invariant is called a symmetry by physicists. Technically, the invariance of the physics is expressed as the invariance of the equations of motion that describe the physics. This is implied by the invariance of the lagrangian which expresses the physics.

Symmetries correspond to conservation of some kind of charge. For example, the invariance of physics under spatial translation (moving from London to Whitby) leads to the conservation of momentum; momentum is the conserved charge.

The physics of the universe is the same on Mondays as it is on Tuesdays. Invariance under temporal translation leads to the conservation of energy. Kettles boil at the same temperature when their spout is pointing north as they do when their spout is pointing west. Invariance under spatial rotation leads to the conservation of angular momentum.

The reader might have noticed that the symmetries we have written of immediately above are translation and rotation in our 4-dimensional space-time. What about symmetries in the quaternion space and in the complex plane?

Some standard mantra:

The standard mantra is that a symmetry is a change in the fields of the lagrangian, that's $\psi(x)\,\&\,\overline{\psi}(x)$ in (9.3), which leaves unchanged the equations of motion derived from that lagrangian. In the case of the electron and the positron, the field in the lagrangian, (9.3), is $\psi(x)$ together with its adjoint (kind of conjugate) $\overline{\psi}(x)$. Suppose we multiply each of these fields by $e^{i\alpha}$ where α is just a real number and not a function of the position in space-time. Of course $e^{i\alpha}$ is just a rotation in the complex plane (not in an $\mathbb{R}^2$ plane of our 4-dimensional space-time). The way we do this is:

$$\psi(x) \to e^{i\alpha}\psi(x) \qquad \overline{\psi}(x) \to \overline{\psi}(x)e^{-i\alpha}$$
$$\partial^{\mu}\psi(x) \to e^{i\alpha}\partial^{\mu}\psi(x) \tag{9.5}$$

We use $e^{-i\alpha}$ on the adjoint field to match the rotational direction, away from or toward the real axis, of the rotation of the field. Since $e^{i\alpha}$ is not a function of the space-time position, it acts as a constant under differentiation.

Putting these into the Dirac lagrangian, (9.3), gives:

$$\mathcal{L}_{Dirac} = i\overline{\psi}(x)e^{-i\alpha}\gamma_{\mu}e^{i\alpha}\partial^{\mu}\psi(x) - m\overline{\psi}(x)e^{-i\alpha}e^{i\alpha}\psi(x)$$
$$= i\overline{\psi}(x)\gamma_{\mu}\partial^{\mu}\psi(x) - m\overline{\psi}(x)\psi(x) \tag{9.6}$$

We see that the lagrangian is unchanged by rotation in the complex plane; this implies that the equations of motion derived from the lagrangian are unchanged by rotation in the complex plane. This is called a $U(1)$ symmetry.

The Lie group $U(1)$ is no more than the rotation group, a circle, in the complex plane, and the reader is entirely correct to think of $U(1)$ as rotation in the complex plane, $\mathbb{C}$. The normal phraseology is to say that $e^{i\alpha}$ is a phase, angle with the real axis, and that $U(1)$ is the group of possible phases.

In short, the physics described by the Dirac lagrangian is invariant under rotation in the complex plane. To reiterate, since the physics of the Dirac lagrangian is unchanged under a $U(1)$ transformation[50], the expression of that physics as the equations of motion derived from the Dirac lagrangian, the Dirac equation, will by unchanged by a $U(1)$ transformation.

There should be a conservation law associated with this $U(1)$ symmetry. There is; it is conservation of electric charge. So, whatever electric charge is, it is associated with rotation in the complex plane. There's something to blow your brains out.

Noether's theorem:
The above idea that a symmetry is associated with a conservation law and a conserved charge has its mathematical proof and expression in a theorem called Noether's theorem. Many folk are of the view that Noether's theorem[51] is the most important theorem of modern physics.

For the time being, for the Dirac lagrangian, Noether's theorem is concerned with the concept of an infinitesimal $U(1)$ transformation; that is an infinitesimal rotation in the complex plane. The understanding behind this is that any rotation in the complex plane can be seen as a large number of infinitesimal rotations[52]. The infinitesimal rotation is:

$$\psi \rightarrow (1+i\alpha)\psi \qquad (9.7)$$

Where α is very small.

We require that the Dirac lagrangian to be unchanged by this infinitesimal rotation in the complex plane. The details of the

---

[50] That's posh talk for rotation in the complex plane.
[51] Named after Emmy Noether (1882-1935).
[52] Your author has a low opinion of this idea.

calculation are available in other literature, and so we give only the result. That result is that the divergence of a current is zero. We have:

$$j^\mu = -e\overline{\psi}\gamma^\mu\psi$$
$$\partial_\mu j^\mu = 0 \tag{9.8}$$

The $-e$ is the electron charge and j^μ is the electromagnetic charge current density. We are using the Einstein summation convention in which:

$$\partial_\mu j^\mu = \frac{\partial}{\partial t} j^t + \frac{\partial}{\partial x} j^x + \frac{\partial}{\partial y} j^y + \frac{\partial}{\partial z} j^z \tag{9.9}$$

If the divergence of a current is zero, then there is a charge given by:

$$Q = \int d^3x \; j^0 \tag{9.10}$$

Where j^0 is the component of the current in the time direction. We see that this charge, Q does not vary with time:

$$\frac{dQ}{dt} = 0 \tag{9.11}$$

This is a conserved charge – it stays the same for all time.

What has happened here is that because rotation in the complex plane, $\mathbb{C}$, does not change the physics expressed in the Dirac lagrangian, a conserved charge exists. By observation and common sense, physicists identify the $U(1)$ conserved charge with electric charge.

Furthermore, rotation in the clockwise direction is associated with negative electric charge as held by the electron and rotation in the anti-clockwise direction is associated with positive electric charge as held by the positron. In general, when working with complex fields, the conjugate of the complex field is taken to be the anti-particle of the particle represented by the field.

Whether the rotation is clockwise or anti-clockwise for negative charge is unimportant. We just arbitrarily associate clockwise rotation with negative charge and thereby associate anti-clockwise rotation with positive charge.

Conservation laws are local:

The reader will be familiar with the law of conservation of energy. The reader probably learnt this law as *"Energy cannot be created or destroyed"*. I know that this law is now modified to allow for conversion of mass into energy, but that is not relevant to this paragraph.

Now, imagine that a kilo-joule of energy suddenly just disappeared from the room in which you are sitting but that, at the very same instance a kilo-joule of energy suddenly appeared on the moon. According to the law of conservation of energy as stated above, is the law of conservation of energy violated by this disappearance and simultaeous reappearance? No. There is as much energy in the universe after these events as before these events. The total amount of energy in the universe never changed.

Properly stated, the law of conservation of energy includes that energy is conserved locally[53]. This means that if energy is going to be transferred from the Earth to the moon, it can be transferred no faster than the speed of light.

This localness of the conserved charge needs to be taken account of in Noether's theorem. You will see us do this shortly when we ask 'from where comes the photon?'.

---

[53] It also has to include "within an inertial reference frame". Otherwise, you could change the total energy of the universe, in violation of the conservation of energy law, by simply accelerating while riding your bike.

A notational difficulty:

The Dirac lagrangian and the nature of the infinitesimal rotation hide a notational mess which is usually glossed over. We take a look at that notational mess now.

Using the notation:

$$a + ib = \begin{bmatrix} a & b \\ -b & a \end{bmatrix}$$

$$\exp\left(\begin{bmatrix} 0 & b \\ -b & 0 \end{bmatrix}\right) = \begin{bmatrix} \cos b & \sin b \\ -\sin b & \cos b \end{bmatrix}$$

(9.12)

We can look at the infinitesimal rotation, (9.7), as:

$$\psi \to \left(\begin{bmatrix} \cos 0 & 0 \\ 0 & \cos 0 \end{bmatrix} + \begin{bmatrix} 0 & \sin\alpha \\ -\sin\alpha & 0 \end{bmatrix}\right)\psi$$

$$= \begin{bmatrix} \cos\alpha & \sin\alpha \\ -\sin\alpha & \cos\alpha \end{bmatrix}\psi$$

(9.13)

We have approximated the $\cos\alpha \approx \cos 0$. The reader is right to ask why we must use infinitesimal rotations rather than just using the proper rotation matrix. Perhaps there is room for improvement in this regard.

Clearly, a $U(1)$ rotation is a complex number. To do the mathematics properly, $\psi(x)$ must be a complex number. The standard mantra is that the Dirac spinor (Dirac field), $\psi(x)$ is four complex numbers in two pairs, and so we have a notational mess here. The notation is obscure, but let us enlighten the reader. It is taken that the $U(1)$ rotation acts on each of the complex numbers separately and that:

$$e^{i\alpha}\psi(x) \equiv e^{i\alpha}\begin{bmatrix} a+ib \\ 0 \\ 0 \\ 0 \end{bmatrix} \& e^{i\alpha}\begin{bmatrix} 0 \\ c+id \\ 0 \\ 0 \end{bmatrix} \& e^{i\alpha}\begin{bmatrix} 0 \\ 0 \\ e+if \\ 0 \end{bmatrix} \& e^{i\alpha}\begin{bmatrix} 0 \\ 0 \\ 0 \\ g+ih \end{bmatrix}$$

(9.14)

We will later take the Dirac field, $\psi(x)$, to be a pair of quaternions which can be separated, decoupled, into two quaternion equations. When we do this, we will have only half the notational mess we had before, but how do we multiply a quaternion (a 4×4 matrix) by a complex number (a 2×2 matrix)? We do it one imaginary variable at a time. We separate the quaternion into three 'double cover' complex numbers and we then disregard the double cover. In notation:

$$
\begin{bmatrix}
a & b & c & d \\
-b & a & d & c \\
-c & d & a & -b \\
d & -c & b & a
\end{bmatrix}
\rightarrow
\begin{bmatrix}
a & b & \sim & \sim \\
-b & a & \sim & \sim \\
\sim & \sim & a & -b \\
\sim & \sim & b & a
\end{bmatrix}
\rightarrow
\begin{bmatrix}
a & b \\
-b & a
\end{bmatrix}
\quad (9.15)
$$

This is not mathematics. This is how we think quaternion space is manifest in our 4-dimensional space-time. There is quaternion space, and, mathematically, quaternion space is separate entirely from our 4-dimensional space-time. Our 4-dimensional space-time is a fabricated space (fabricated from the A_3 algebras) and not an algebraic mathematical structure – not a division algebra. We cannot expect the rules of mathematics which apply within a division algebra like the quaternions to apply within a fabricated space or within the interaction of two different division algebras. We do not see double cover quaternion rotation in our 4-dimensional space-time because our 4-dimensional space-time holds only 2-dimensional single cover rotations. We presume a quaternion rotation will appear in our 4-dimensional space-time as a 2-dimensional rotation.

Perhaps the notation of the Dirac lagrangian is a mess. Perhaps the reader will feel the above disentanglement of the notation is contrived. Perhaps the reader is correct, but the Dirac equation is central to quantum field theory; we cannot simply reject it until we have something with which to replace it. This proposed replacement will be presented in a later chapter.

From where comes the photon?:

In deducing a connection between electric charge and rotation in the complex plane from the invariance of the Dirac lagrangian under that rotation, it is assumed that the rotation was independent of position in space-time. This is called a global symmetry because the rotation is the same at all points in our 4-dimensional space-time universe.

This is to say that, assuming I can do it, if I suddenly rotate in the complex plane through, say 30^0, here in my study, the whole universe instantly rotates through 30^0 to match. But such instantaneous change throughout the universe is not possible in our 4-dimensional space-time. More realistic is a sudden rotational change in my study radiating outward at the speed of light into the rest of the universe. In this realistic scenario, the rotation depends upon the position in space-time – it is 30^0 in my study, but, for the next 2.5 million years, it will be zero degrees in the Andromeda galaxy. What happens if the complex plane rotation, the $U(1)$ transformation to use the standard terminology, depends upon the space-time position?

Consider:

$$\psi(x) \to e^{i\alpha(x)}\psi(x)$$
$$\overline{\psi}(x) \to \overline{\psi}(x)e^{-i\alpha(x)}$$

(9.16)

Notice the angle parameter, $\alpha(x)$, is a function of position in space-time. The rightmost and leftmost two terms of the Dirac lagrangian, (9.3), are invariant under this transformation (rotation), but the differential term is not invariant under this transformation.

Clearly, physics should be invariant under rotation in the complex plane. Hm! really, it is not at all clear that this must be the case; in fact, this is quite an amazing assumption, but let us presume it is the case. We have:

$$\partial_\mu \psi \to \partial_\mu \left(e^{i\alpha(x)}\psi\right) = e^{i\alpha(x)}\partial_\mu \psi + i\partial_\mu \alpha(x)e^{i\alpha(x)}\psi$$

(9.17)

We have an extra term due to the dependence of the phase $\alpha(x)$ on the position in 4-dimensional space-time. In order to have the Dirac lagrangian invariant under local rotation in the complex plane, that is local $U(1)$ invariance, we need to subtract the term $i\partial_\mu\alpha(x)e^{i\alpha(x)}\psi$ when we differentiate. We replace the normal differentiation with a covariant differentiation, D_μ, such that:

$$D_\mu\psi = \partial_\mu\psi - i\partial_\mu\alpha(x)e^{i\alpha(x)}\psi \qquad (9.18)$$

This gives invariance of the differential term in the Dirac Lagrangian. We have rescued the invariance, but we have been forced to introduce a new type of differentiation involving a new field. We put:

$$e_c A_\mu = \partial_\mu\alpha(x) \qquad (9.19)$$

wherein e_c is the charge of the electron and A_μ is called a gauge field. It so happens that A_μ has the properties of the photon field.

Thus, we see that the need to have the Dirac lagrangian (it could be any lagrangian with a differential term in it) invariant under under local rotation in the complex plane 'causes' the photon to exist.

It is as if the photon communicates the angle in the complex plane, the $U(1)$ phase, between each point in 4-dimensional space-time.

The electromagnetic force briefly:
The force of gravity is described by the general theory of relativity. According to the general theory of relativity, all objects move along geodesics in space-time. Basically, all objects remain in their state of motion and simply move in an 'unaccellerated straight line', a geodesic, through space-time. These 'unaccellerated straight lines' appear to be accellerated because 4-dimensional space-time can be curved. Thus, the gravitational force is seen by general relativity as being an 'illusion' caused by 4-dimensional space-time being curved.

How about the electromagnetic force? Is the electromagnetic force somehow due to a kind of 'curved electromagnetic space'. It might be. The mathematics of expectation spaces lead to two spaces. These two expectation spaces are our 4-dimensional space-time and a quaternion expectation space. Each of these spaces has within it an affine connection (a sense of what it means for two lines to be parallel). Perhaps the quaternion expectation space could be curved and perhaps this curvature is what we call an electro-magnetic field.

It is the case that the photon field is a measure of how 'out of parallel' the phase of the electron field is from point to point in space-time, and we can associate this with a kind of curvature.

Well, the electromagnetic force might be due to some kind of curved quaternion expectation space, but that is not the direct way quantum field theory, QFT, sees it. The quantum electro-dynamics, QED, of the electromagnetic force is seen by QFT as being 'caused' by photons. The idea is that two charged bodies repel, or attract, each other because photons, which have momentum, are 'flying' back and forth between the two charged bodies. This will be looked at in more detail later; for now we give an example, two electrons repel each other because they recoil from emitting and absorbing the momentum of photons which leave them or hit them. So, rather than a smooth bending of the electron's trajectories, we have a sharp kink in their trajectories as the photons are emitted or absorbed.

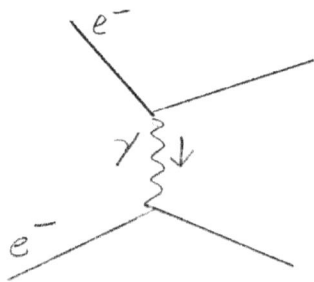

The strength of the electromagnetic force is then equal to the probability of an electron emitting or absorbing a photon. Hang on! what was that last sentence? The strength of the electromagnetic force is equal to the probability of an electron emitting or absorbing a photon. Yes, this is exactly how QFT sees the fundamental strength of the electromagnetic field, the value of the fine structure constant, to be determined. The fine structure constant is a measure of the probability that an electron will emit or absorb a photon. This is a measure of the strength of the electromagnetic force.

Actually, with a little thought, we realise that the strength of the electromagentic field is also a function of what the probable momentum of the probably emitted photon will be, but that's just details. It is the idea that matters. Strength of electromagnetic force is probability of photon emission by an electron.[54]

The electromagnetic force is called the electromagnetic interation in the jargon of QFT.

Summary:

This chapter has been quite shocking. We have avoided the mathematical details to avoid obscuring the concepts. The Dirac equation is conventionally taken to describe all fermions. The requirement that the physics of the Dirac lagrangian which leads to the Dirac equation be invariant under rotation in the complex plane - $U(1)$ invariance – has led to the existence of electric charge and the existence of the photon. This is the standard mantra of quantum field theory. Notice how the photon, light, and electric charge are closely tied to each other.

The derivation of electric charge and of the photon from this one simple symmetry is one of the great triumphs of quantum field theory. Doubtless the reader in thinking, "What about rotation in quaternion space?". Let us not trip over ourselves. Let us take time to think. The existence of electromagnetic phenomena such as light and electric charge are intimately connected to the existence of the complex plane, $\mathbb{C}$. Can we say that the existence of light proves the existence of the complex plane space? It seems so.

---

[54] You may use this great knowledge to chat up theology students. Of course, you will have to wait until you go to heaven to get your reward.

Before ending this chapter, we will just reiterate something. The strength of the electromagnetic interaction is the probability of an electron emitting or absorbing a photon.

Chapter 10

The Uncertainty Principle

A central part of humankind's understanding of the physical universe is the uncertainty principle. This principle is often called the Heisenberg uncertainty principle in honor of Werner Heisenberg (1901-1976).

In mathematical terms, the uncertainty principle is that, given any two quantum mechanical operators, $\{A, B\}$, we have:

$$\Delta A \Delta B \geq \frac{\langle AB - BA \rangle}{2i} \tag{10.1}$$

wherein ΔA is the uncertainty in the operator A and ΔB is the uncertainty in the operator B. An example is the two operators for position, X, and for momentum, P. Any quantum mechanics text will tell you that $XP - PX = i\hbar$. This leads to:

$$\Delta X \Delta P \geq \frac{\hbar}{2} \tag{10.2}$$

Another example is:

$$\Delta E \Delta T \geq \frac{\hbar}{2} \tag{10.3}$$

The first of these, (10.2), says that it is impossible for a particle to simultaneously have a definite spatial position in 4-dimensional space-time and a definite spatial momentum in 4-dimensional space-time. The particle can have a very accurately defined spatial position in 4-dimensional space-time, but, if it does have an accurately defined spatial position, then it must have a very inaccurately defined spatial momentum. Similarly but oppositely, the particle can have a very accurately defined spatial momentum in 4-dimensional space-time, but, if it does have an accurately defined spatial momentum, then it must have a very inaccurately defined spatial position.

The second of these, (10.3), says that it is impossible for a particle to have a definite energy (temporal momentum) in 4-dimensional space-time and a definite temporal position in 4-dimensional space-time simultaneously. A temporal position is the same as a spatial position but upon the time axis of our 4-dimensional space-time rather than upon a spatial axis of our 4-dimensional space-time. An uncertainty is spatial position is easier to imagine than an uncertainty in time position, but, since we live in space-time and the time axis is equivalent to the space axis, if we can have uncertainty in spatial position, we can have uncertainty in temporal position.

Uncertainty in time position – rather mind blowing.

Clearly, to an observer in 4-dimensional space-time, the above two uncertainty relations, (10.2) & (10.3), are the same thing.

In the view of QFT, the consequences of the uncertainty principle are of enormity within our physical universe. We would very much like to understand why there is uncertainty in spatial or temporal position and uncertainty in momentum and energy.

Attempts to explain the uncertainty principle:
Over the past century, different explanations of the uncertainty principle have been presented by different people; none of these explanations have been really convincing explanations. Still today, there is no universally accepted explanation of the uncertainty principle in particle physics.

We will offer different explanations of why we have an uncertainty principle in our universe. The reader might find none of these explanations convincing.

Uncertainty in classical physics:

It is universally accepted that there is uncertainty in classical wave phenomena. If we allow that particles are wavelike, then the classical uncertainty of wave phenomena automatically becomes part of particle physics.

Every classical physicist who works with classical waves knows that it is impossible to define a unique wavelength for a short wave train. A short wave train is a small number of wave crests:

Basic quantum mechanics associates both energy and momentum with wave mathmatical properties in the equations:

$$E = \hbar\omega$$
$$p = \hbar k$$

(10.4)

wherein ω is the wave frequency and k is the wave number. We see that the wavelength, the distance between wave-crests, corresponds to the momentum. Unless, the wave train is infinitely long, we have an uncertainty in the wavelength (momentum). It is a result of classical physics that the product of the length of the wave-train and the uncertainty in the wave-number associated with that wave train is at least 2π. This is nothing to do with quantum mechanics; this is a purely classical phenomenon.

If we are going to associate a particle's momentum with the wavelength of that particle when it is in wave form, then we are going to have some form of uncertainty in the product of the particle's momentum and a measure of the length of the wave-train. In short, if particles have wave-like properties, then we will have the uncertainty principle.

Uncertainty from Fourier transforms:
Within the complex numbers, $\mathbb{C}$, we have two trigonometric functions which are the cosine function and the sine function. In spite of our mathematical eduction teaching us to use these functions to calculate the height of distant church towers, these functions do not exist outside of the complex numbers. The quaternion trigonometric functions are similar to these functions, but they are not the same.

The sine and cosine functions are wave functions. Every mathematical expression of a wave uses one or both of these functions. We thus conclude that waves exist in only the complex numbers. Well, that is a little premature. Our 4-dimensional space-time is a fabrication of six 2-dimensional spinor spaces. Three of these six spinor spaces are the complex planes, and so waves can exist in our 4-dimensional space-time. Further, quaternion space has a kind of wave functions for its trigonometric functions.

There is an area of mathematics called Fourier theory which deals exclusively with the cosine and sine functions. Fourier theory has within it the concept of position space wave function connected via a Fourier transform to a momentum space wave function. Within the mathematics of Fourier theory, there is an uncertainty relation connecting the momentum space to the position space which corresponds to the uncertainty principle[55].

Thus it is that the concept of waves, expressed as the trigonometric functions of the complex plane, $\mathbb{C}$, have an inherent uncertainty between position and momentum about them.

Uncertainty from standard deviation:
When physicists predict the result of an experiment using quantum mechanics, they actually predict an average result known as an expectation value. Because the outcome of a given series of events is

---

[55] See Davis McMahon Quantum Mechanics DeMystified pg 54

not determined but has a probability associated with it, we have to take averages, which we call expectation values.

Indeed, it seems that the whole of our classical universe is one big expectation universe; that is one big average of quantum physics. An example of this is the 4-dimensional space-time of our universe which is fabricated around an expectation distance function which is the average of the six A_3 algebra distance functions.

Associated with the statisical mathematics of taking an average is a statisical concept called the standard deviation which measures the 'accuracy' of the measured average. When applied to quantum mechanics, the standard deviation becomes the uncertainty principle[56].

Uncertainty from quaternion space?:
We showed above that the mismatch of the distance function within quaternion space and the distance function of our 4-dimensional space-time are such that an electron in quaternion space can have spatial extent in our 4-dimensional space-time only instantly or it can have temporal extent only if it has no spatial extent. The velocity, and hence the momentum of an electron is meaningful only if there is both spatial extent and temporal extent associated with the electron – metre per second. Absence of either temporal extent or spatial extent makes velocity, and hence momentum, meaningless.

Since electrons have momentum, it seems there must be some uncertainty in the way the quaternion distance function compares with the 4-dimensional space-time distance function.

Perhaps it simply is that no-one, not even the universe, can measure distance more accurately than a wavelength of light of a particular frequency and that frequency is connected to the amount of momentum of an electron. Not very convincing.

---

[56] See David McMahon Quantum Mechanics DeMystified pg 185

Earlier attempts to explain the uncertainty principle:

In many of the texts written more than twenty years ago, the reader might find an explanation of the uncertainty principle first presented by Werner Heisenberg. The idea is that we measure the position of, say, an electron by bouncing a photon of light off the electron. When we make this measurement, we must change the position of the electron because we have just hit the electron with a photon. A similar argument applies to the momentum of the electron being changed by being hit by a photon. This neccessity to change the position and momentum of what we measure whenever we measure it is presented as being an uncertainty in our measurement; this is an uncertainty in our knowledge of the universe not an uncertainty in the physics of the universe.

This explanation of the uncertainty principle was so widely presented because no other prominent physicist offered a different explanation in its place. This explanation of the uncertainty principle is now rarely presented because it is felt that the uncertainty principle should be intrinsic to reality rather than intrinsic to human knowledge of reality. Much of what happens in the universe happens without a human observer yet necessarily depends upon the uncertainty principle.

Summary:

We do not properly understand the uncertainty principle.

We are quite confident that uncertainty is within the universe and not within our knowledge of the universe.

In the next chapter, we will meet the profound consequences of the uncertainty principle.

Addendum:

We have shown elsewhere[57] that the scaling parameter, λ, within the complex numbers (the parameter which sets the relative scales of the real axis and the complex axis) is such that, at least within the momentum operator of quantum mechanics, we have:

$$\hbar = \frac{1}{\lambda} \tag{10.5}$$

Is this a clue to anything? It almost certainly is a clue to something, but what?

---

[57] See: Dennis Morris The Physics of Empty Space circa pg 90

Chapter 11

Virtual Particles

We reiterate that a space such as quaternion space which has no time within it does not have meaningful temporal conservation laws.

An utterly central concept of quantum field theory is the concept of the virtual particle existing within a virtual process. The concept of the virtual particle is entirely based upon the uncertainty principle. The reader is reminded of the energy-time uncertainty relation:

$$\Delta E \Delta T \geq \frac{\hbar}{2} \qquad (11.1)$$

Time exists in our 4-dimensional space-time, and so the above statement, (11.1), is about something in our 4-dimensional space-time.

The $\geq$ in (11.1) is really the statement that human knowledge can be more inaccurate than reality. We really ought to stop considering human knowledge and consider instead the universe. The relation (11.1) should have an equals sign rather than a $\geq$ sign:

$$\Delta E \Delta T = \frac{\hbar}{2} \qquad (11.2)$$

This relation is nothing to do with human knowledge. This relation is believed to express a basic fact about the universe. If the uncertainty principle, (11.2), is not a basic fact of the universe, then then whole of quantum field theory is wrong.

Here's the rub:
The relation (11.2) allows an amount of energy, ΔE, to pop into existence in our 4-dimensional space-time from nowhere for a small amount of time, ΔT, provided the product $\Delta E \Delta T$ is less than or

equal to $\frac{\hbar}{2}$. This might seem like a direct violation of the conservation of energy. It is a direct violation of the conservation of energy, but it can never be measured. We are not writing of uncertainty in human knowledge; we writing about the uncertainty in existence within our 4-dimensional space-time.

The close to instantaneous dynamics of individual particles like photons and electrons are not constrained by energy conservation or by momentum conservation; only the result of an enduring process is constrained by these conservation laws. This absence of conservation contraints allows an electron to emit a photon or a photon to convert into an electron and a positron over microscopic timescales provided that energy and momentum are conserved over macroscopic timescales. It is as if energy conservation and momentum conservation are constraints over the dynamics of our 4-dimensional space-time but are not constraints over any dynamics in spaces, like quaternion space[58], other than our 4-dimensional space-time.

These 'illegal' violations of the conservation of energy law and the conservation of momentum law are called virtual particles or virtual processes, and they are intermediate states within elementary particle interactions. We do calculations in QFT like Feynman diagrams by taking account of every possible virtual process and summing the probabilities of them happening. The reader might think that, since we do not observe these virtual processes, they do not happen. Perhaps the reader is correct, but they really exist in the calculations of QFT; without the virtual processes, we would by unable to do the vast majority of calculations that are the bedrock of QFT.

Technically, any product of energy and time less than $\frac{\hbar}{2}$ can 'pop' into our 4-dimensional space-time, but it is meaningless to consider the product of energy and time to be 'less than' $\frac{\hbar}{2}$ because, by virtue

---

[58] Quaternion space is without time, and so an uncertainty relation involving time would be meaningless in quaternion space.

of the uncertainty relation itself, the energy can never be measured with sufficient accuracy to distinguish between the $\frac{\hbar}{2}$ and less than $\frac{\hbar}{2}$. Thus, we might as well take the product $\Delta E \Delta T$ to equal $\frac{\hbar}{2}$.

Based on the fact that the uncertainty relation makes it possible for an amount of energy to pop into existence in our 4-dimensional space-time from nowhere for a small amount of time, quantum field theory assumes that the energy does pop into existence in our 4-dimensional space-time for that small amount of time. This means that, from nowhere, little bits of energy are popping into existence within our 4-dimensional space-time, enduring for a short time, and then disappearing out of our 4-dimensional space-time.

"But I thought energy was conserved", I hear the reader cry. At school, the reader learnt, "*Energy is niether created nor destroyed locally within an inertial reference frame*", but there is a bit missing. Energy is conserved as accurately as it can be measured. A fundamental inaccuracy is imposed by the uncertainty principle. Within the limits of that inaccuracy, energy can pop out of nowhere and energy can disappear back into nowhere. Such popping in and out of existence in our 4-dimensional space-time is called a quantum fluctuation. QFT assumes the whole of our empty 4-dimensional space-time is a seething 'froth' of bits of energy popping in and out of existence. We call the 'bits of energy' virtual particles.

That's correct. The QFT view of the vacuum of empty space is that this vacuum is filled with bits of energy, virtual particles, popping in and out of existence in our 4-dimensional space-time. Obviously, no sensible person would believe such a fantastic notion. Remarkably, there is experimental evidence to support this virtual particle understanding – see the Casimir effect below. Fantastic as the notion is, it is backed by experimental evidence.

Energy needs to be associated with some sort of physical object, a particle or a wave. There are different types of particle. Perhaps the energy pops into our 4-dimensional space-time as a photon or perhaps as an positron – no it cannot be a positron because electric charge is conserved and there is no uncertainty relation involving electric charge[59]. Of course, electric charge is associated, by Noether's theorem, with the complex plane rather than with our 4-dimensional space-time. However, a photon with sufficient energy to 'decay' into both an electron and a positron can pop into our 4-dimensional space-time, and this does not violate conservation of electric charge. The photon, having decayed into an electron and a positron, can then be reformed by the electron and the positron recombining. We have a picture in which time runs from left to right[60]:

Of course, this process is not limited to only photons, positrons and electrons. A photon, or any other particle which is its own anti-particle, bosons are their own anti-particles, can pop into our 4-dimensional space-time with sufficient energy to create, say, a tau particle and an anti-tau particle which then recombine to form a photon which then pops out of our 4-dimensional space-time.

There is a time constraint. It takes a lot of energy to form an electron and a positron; thus, under the uncertainty relation, (11.2), these two particles can exist for only a short time. It takes much more energy to form a tau particle and an anti-tau particle, and so these particles can exist for only a much shorter time than the electron positron pair.

---

[59] Quite why there is no uncertainty relation regarding electric charge is a good question.
[60] Within the circles of art aficionados, this is called this pop art.

The Casimir effect:

When we look at the Schrödinger equation later, we will see that a free electron is taken to be a superposition of waves of different frequencies. Quantum physics sees particles like the electron to be a 'sum of waves' of different frequencies. Some of these waves which sum to form the particle have wavelengths greater than, say, one metre, but most have wavelengths of less than one metre. Since real particles are of this 'sum of waves' form, then virtual particles ought to be of the same form.

The presence of these virtual particles creates a pressure in the vacuum of our 4-dimensional space just as we would feel an air pressure on the back of our hand if someone suddenly popped a lot of air molecules into existence close to the back of our hand.

In 1948, it was pointed out by Hendrik Casimir (1909-2000) that if two perfectly flat metal plates were hanging parrallel to each other, only waves of wavelength equal to or less than the distance between the plates could fit between the plates. Now, a virtual particle which is comprised of only the short wavelength waves has less energy than a particle comprised of all wavelength waves. This less energy ought to mean less pressure is exerted by the 'only short wavelength' virtual particle.

The less pressure exerted by the 'sum of only short wavelength waves' virtual particles popping into existence means that two parrallel metal plate hanging close together will experience a pressure of the vacuum force pushing them together because the vacuum pressure between the metal plates is less than the normal vacuum pressure around the metal plates. Calculations show the actual force acting on the plates will vary with the fourth power of the distance between the plates.

This attraction of two parallel metal plates to each other due to the pressure of virtual particles popping into our 4-dimensional space-time is known as the Casimir effect.

If it should be found that two parrallel metal plates are attracted towards each other when hanging close to each other in a vacuum, then we have good evidence of the existence of virtual particles for there is no other known mechanism that would cause such attraction.

Let us all genuflect to experimental physicists for, in 1957[61] and in 1958[62], the Casimir effect was experimentally verified. Even physicists admire physicists.

It was verified with greater accuracy in 1997[63], 1998[64], and 2002[65]. "But I cannot believe in virtual particles", I hear the reader's thoughts. Well, your author and others like him also struggle to accept these ideas, but the experiments, damned experiments… How might we escape these damned experiments?

Perhaps the Casimir effect has been misinterpreted and there is some other reason why the parallel close metal plates attract each other, but no-one thinks this is the case. Thus, the presence of virtual particles popping in to and out of our 4-dimensional space-time has been experimentally verified. Now there's something that will impress a sociology student.

On mass-shell:
Let us consider the annihilation of a virtual positron and a virtual electron into a virtual photon. The energy of the virtual positron and the virtual electron is:

$$E_{e^+e^-} = \left(m_{e^+}^2 c^4 + p_{e^+}^2 c^2\right)^{\frac{1}{2}} + \left(m_{e^-}^2 c^4 + p_{e^-}^2 c^2\right)^{\frac{1}{2}} \qquad (11.3)$$

[61] Spamaay M. J. (1957) Attractive forces between flat plates – Nature 180 (4581) 334

[62] Spamaay M. J. (1958) Measurements of attractive forces between flat plates Physica 24 (6-10) 751

[63] Lamoreaux S. K. (1997) Demonstration of the Casimir force in the 0.6 to 6 µM range Phys. Rev. Lett. 78 : 5

[64] Mohideen U. Roy. Anushree (1998) Precision measurement of the Casimir force from 0.1 to 0.9 µM Phys. Rev. Lett. 81(21) : 4549

[65] Bressi G. Carugno G. Onofrio R. Ruoso G. (2002) Measurement of the Casimir force between parallel metallic surfaces Phys. Rev. Lett. 88(4) 041804

The energy of the virtual photon is:

$$E_\gamma = p_\gamma c \tag{11.4}$$

Thus, because of the rest masses of the electron and the positron, we do not have a balance between both the energy and the momentum. We do not have:

$$E_{e^+e^-} = E_\gamma \qquad \& \qquad p_\gamma = p_{e^+} + p_{e^-} \tag{11.5}$$

There is violation of the energy and momentum conservation laws here. These conservation laws are intrinsic to the dynamics of our 4-dimensional space-time. Thus, it seems, these virtual processes are happening outside of our 4-dimensional space-time.

Virtual particles are said to be *"off mass-shell"* because they violate the relation $E^2 = p^2c^2 + m^2c^4$. It seems that this relation is a 4-dimensional space-time relation. 'Normal' particles, that is particles which do not violate this relation are said to be *"on mass-shell"*.

The electromagnetic force:

Although empty space is a seething mass of photons popping in and out of our 4-dimensional space-time. It seems that the 'bit of empty space' which holds an electron, or other charged particle, is a centre for such virtual photons. Rather than just sitting there and doing nothing, we view an electron as constantly emitting and absorbing virtual photons.

Repulsion of two electrons is then easy to understand as one electron emits a photon which is absorbed by another electron. The momentum of the photon pushes the electrons away from each other.

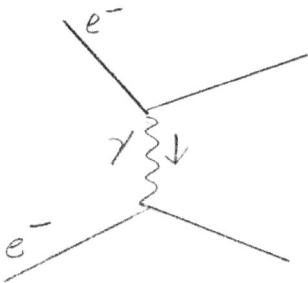

But what about atraction between an electron and a positron? Attraction between a positron and an electron requires that a virtual photon be emitted in the opposite direction to the above diagram and absorbed in the opposite direction to the above diagram. We need:

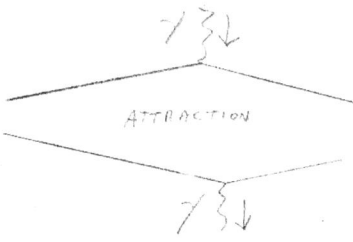

These are virtual photons. There is uncertainty in their spatial position, ΔX, this means it is possible for a photon to leave at the bottom of the picture but to later reappear at the top of the picture so that the momentum fits properly to cause attraction. Do you believe that? This is the standard explanation given for attraction between a positron and an electron according to quantum field theory. I think I prefer the directionally unbalanced emmission of virtual photons explanation.

Charge screening:

We have a view of the electron as constantly emitting and absorbing virtual photons. Between being emitted and being reabsorbed, these virtual photons can 'decay' into a virtual positron and a virtual electron which then recombine to form the virtual photon before it is reabsorbed. We thus have a view of the electron as a point particle surrounded by a swarm of photons and electron-positron pairs. We ignore the photons for the purposes of this paragraph.

While the virtual electron positron pair exist, the electric charge of the real electron repels the virtual electron but attracts the virtual positron. This leads to a polarised virtual screen of charge surrounding the real electron. More great art is now presented:

The swarm of virtual electron-positron pairs flicking in and out of our space-time effectively gives the electron a spatial size.

We now have several ways of asking the same question:

i)   What is the charge of the electron?
ii)   What is the strength of the electromagnetic interaction?
iii)   What is the probability of an electron emmitting or absorbing a photon?

The strength of the electromagnetic interaction, the charge of the electron, depends upon the distance from the electron at which it is measured. The presence of the polarised electron-positron pairs affects the measured charge of the electron, the measured electromagnetic interaction, when we are close to the electron. When we are far from the electron, the measured value of the electromagnetic interaction, which we call the fine structure constant or the electromagnetic coupling constant, is the asymptotic value of $\alpha = \dfrac{1}{137}$, but, very close to the electron, the measured value of the electromagnetic interaction increases dramatically. More art:

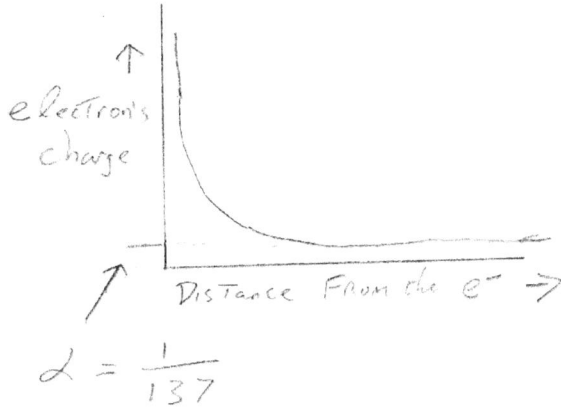

We see that the value of the 'physical contant' we call the fine structure constant varies with distance from the electron. The fine structure is also known as the electromagnetic coupling constant; it is a 'running coupling constant' and the reader might often see this phrase attached to it. In general, the strength of forces, gravity, the weak force, and the strong force, as well as the electromagnetic force are each measured by a coupling constant which is just a single real number.

A digression:

This book is about the electron, but we will digress slightly. Within quantum chromodynamics, QCD, the 'charge screening' effect works oppositely to the QED charge screening shown above. In QCD, the gluons interact with virtual gluons in such a way that the colour charge of the strong force between quarks weakens as we get closer to the quark but strengthens as we get further from the quark. This effect is called 'asymtotic freedom'.

Chapter 12

Neutrinos and the Weak Force

Beta decay:

Within an atomic nucleus, a proton will sometimes change into neutron and a neutron will sometimes change into a proton. This allows the nucleus to change to a stable optimum ratio of protons to neutrons. Such change is a kind of radioactive decay called beta decay because the change involves the nucleus emmiting what was origionally called a beta particle. Today, we know the emmited beta particles are either electrons or positrons. An example of beta decay is:

$$C_6^{14} \rightarrow N_7^{14} + e^- + \overline{v_e} \qquad (12.1)$$

wherein C_6^{14} is a carbon nucleus, N_7^{14} is a nitrogen nuceus, e^- is an electron, and $\overline{v_e}$ is a particle known as an anti-electron-neutrino. The symbol for a neutrino is v_i where i is either an e for the electron neutrino or μ for a muon neutrino or τ for a tau neutrino. Each of these symbols with a bar over it represents the corresponding anti-neutrino. Other examples of beta decay within an atomic nucleus are:

$$Co^{60} \rightarrow Ni^{60} + e^- + \overline{v_e}$$
$$O^{14} \rightarrow N^{14} + e^+ + v_e \qquad (12.2)$$

A positron is symbolised by e^+, and an electron neutrino is symbolised by v_e.

Outside of an atomic nucleus, a neutron can decay into a proton by beta decay but a proton cannot decay into a neutron; because the neutron is (slightly) more massive than the proton, there is insufficient energy outside of an atomic nucleus to allow the proton decay into a neutron. Outside of an atomic nucleus, we have:

$$n \rightarrow p^+ + e^- + \overline{v}_e \qquad (12.3)$$

This process, (12.3), has a half-life of 920 seconds.

The proton mass is 1836 electron masses; the neutron is 1839 electron masses. If these masses of the proton and the neutron were the other way around, then a proton would decay into a neutron, and there would be no free protons in the universe. No free protons means no hydrogen, and this means no stars, and this means no chemical elements, and this means no molecules, and this means no anything familiar to us, and this means no us. If the electron mass was such that it was greater than the mass difference of the proton and the neutron, there would be no neutron decay into a proton. This would mean the universe was full of neutrons left over from the big bang at the start of the universe, and this would mean any molecules would be smashed to bits by all the neutrons, and this would mean no us. No-one understands why these particles have the masses they do. It is most sobering to realise how precariously the nature of our existence depends upon the relative masses of these three particles.

Radioactive decay happens in connection with one of the three basic non-gravitational forces of nature which are the strong nuclear force, the electromagnetic force, or the weak nuclear force. Particles which decay by the strong nuclear force, usually called just the strong force, typically have half-lives of 10^{-23} seconds while particles which decay by the electromagnetic force typically have half-lives of 10^{-16} seconds. Particles which decay by the weak nuclear force, usually called just the weak force, are markedly longer lived with half-lives much longer than those associated with either the strong force or the electromagnetic force.

The decay of the neutron is via the weak force; if the weak force did not exist, the neutron would be stable (that means a half-life in excess of 10^{30} years).

In a more fundamental view, the decay of a neutron into a proton is a down quark with electric charge $-\dfrac{1}{3}$ becoming an up quark with electric charge $+\dfrac{2}{3}$. We have:

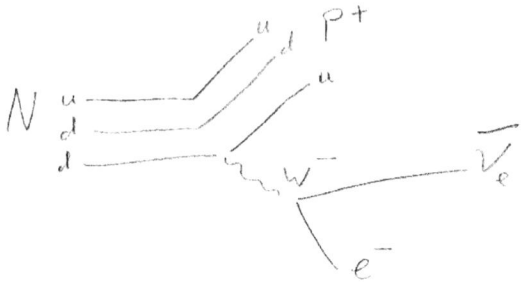

With the aid of the artistically meritorious diagram above, we see that beta decay involves a W^- boson. The three $\{W^\pm, Z^0\}$ bosons are the bosons of the weak force; all decays involving one or more of these weak force bosons is a weak interaction. These weak force bosons, $\{W^\pm, Z^0\}$, have never been seen; their existence is inferred from the observation of the particles into which they decay. Clearly, a quark changing from a down quark into an up quark by emitting a W^- boson is a weak force process.

Neutrinos interact through only the weak force. If a neutrino is involved, the weak force is involved. However, sometimes the weak force is involved but there are no neutrinos – see decay of kaons below.

Beta decay was discovered by Henri Becquerel (1852-1908) in 1896, but it was not understood until the 1930's. Although electrons had been known since Thompson's discovery of them in 1896, the anti-neutrino/neutrino was first postulated to exist by Wolfgang Pauli (1900-1958) in 1930 as being necessary in beta decay to conserve energy, momentum, and spin angular momentum. The postulated particle was named the neutrino by Enrico Fermi (1901-1954) in 1931. In 1942, Wang Ganchang proposed using beta capture to detect neutrinos[66]. Neutrinos, actually anti-neutrinos, were first seen experimentally in 1956 in the Cowan-Reines neutrino experiment done by Clyde Cowan (1919-1974), Frederick Reines (1918-1998), F. B. Harrison, H. W. Kruse and A. D. McGuire.[67] In 1995, Cowan and Reines were awarded the Nobel Prize for this experiment.

In 1962 Leon M Lederman (1922-), Melvin Schwartz (1932-2006) and Jack Strinberger (1921-) first detected the muon neutrino thereby showing that more than one type of neutrino exists. For this they were awarded the 1988 Nobel Prize. The tau particle was discovered in 1975 at the Stanford Linear Accellerator; with this discovery came the expectation that a tau neutrino exists. The tau neutrino was discovered in 2000 by the DONUT collaboration at Fermilab.

What are neutrinos?:

Neutrinos, and anti-neutrinos, have zero electric charge. Associated with zero electric charge, neutrinos have zero magnetic dipole moment. They have spin one half – this means neutrinos, and anti-neutrinos, are fermions. They have lepton number, although things are not that simple in this regard. They also carry a kind of charge (not electric charge) called weak hypercharge of which they have one unit. Weak hypercharge is like electric charge except it is not electric charge; it is weak hypercharge. Like photons, neutrinos have energy; the amount of energy associated with each neutrino differs from neutrino to neutrino, as it does with photons.

For many years, it was assumed that neutrinos were massless, but there is now evidence that they have a very small mass; there is also evidence that neutrinos are massless – perhaps we are misinterpreting something. As with lepton number, in the case of neutrinos, mass is not so simple.

---

[66] K. C. Wang (1942) A Suggestion on the detection of the Neutrino Physical Review 61 (1-2) 97

[67] C. L. Cowan Jr., F. Reines, F. B. Harrison, H. W Kruse et al (1956) Detection of the Free Neutrino a Confirmation Science 124 (3212) 103.4

It is established that there are three types of neutrino corresponding to the three types of electron-like fermions which are the electron, the muon and the tau particle. So it is that we have the electron neutrino, v_e, the muon neutrino, v_μ, and the tau neutrino, v_τ, and the corresponding anti-neutrinos. Each generation of particles are distinguished from each other in only that their masses are different. It therefore seems that the minute masses of the three types of neutrino must be different, but the actual mass of a single type of neutrino has never been measured. From cosmological considerations, it seems that the sum of the masses of the three different types of neutrino is:

$$m_{v_e} + m_{v_\mu} + m_{v_\tau} = 0.32 \pm 0.081 \ \frac{eV}{c^2} \qquad (12.4)$$

There are other ways of estimating neutrino mass which give similar very small results.

Remarkably, and seemingly in conflict with the assertion that neutrinos have mass, neutrinos seem to always move at the speed of light. Comparison measurements of the velocity of neutrinos to the velocity of light from the 1987A supernova[68] give the velocity[69] of neutrinos to be the same as the velocity[70] of light within a margin of error $\frac{|v-c|}{c} < 2 \times 10^{-9}$. Lorentz violating framework experiments have reduced this margin of error to $\frac{|v-c|}{c} < 5.6 \times 10^{-19}$.[71] There is much evidence that neutrios have no mass; there is evidence that

---

[68] Slodolsky Leo (1988) The Speed of Light and the Speed of Neutrinos Physics Letters B 210(3) 353-354
[69] Hirata et al (1987) Observation of a neutrino burst from the supernova SN1987A Physical Review Letters 58(14) 1490-1493
[70] Longo Michael J (1987) Tests of Relativity from SN1987A Physical Review D 236(10) 3276-3277
[71] Steckler, Floyd W (2014) Constraining Superluminal Electron and Neutrino Velocities using the 2010 Crab Nebula Flare and the Ice Cube PeV Neutrino Events Astroparticle Physics 56 16-18 arXiv 1306-6095

neutrinos have a very small mass; we do not understand neutrino mass; well not until a later chapter perhaps.

Creation of neutrinos:
When-ever an anti-electron-neutrino is created, an electron is also created. When-ever an electron neutrino is created, a positron is also created. We might think of the electron and the anti-neutrino as birth twins except that there are electromagnetic processes (not weak processes) which create electrons or positrons but do not simultaneously create anti-neutrinos and neutrinos. Such a process is the decay of a photon into an electron and a positron. Similarly, when muon neutrinos or tau neutrinos are created, a muon or a tau particle respectively are simultaneously created.

Processes involving beta decay in the core of the sun produce copious electron neutrinos. There are 6.5×10^{14} neutrinos per second from the sun passing through each square metre of the Earth.

Other weak decays:
Although we have introduced the neutrino through beta decay, there are other types of particle decay which involve neutrinos. In fact, all hadrons and all leptons 'feel' the weak force, the weak interaction as it is often called. Examples are:

$$\pi^- \rightarrow \mu^- \overline{\nu}_\mu$$
$$\mu^- \rightarrow e^- \overline{\nu}_e \nu_\mu \tag{12.5}$$

The pion, π^-, has a half-life of 2.6×10^{-8} seconds. The muon, μ^-, has a half-life of 2.2×10^{-6} seconds. These half-lives are a billion times longer than the half-lives associated with the electromagnetic interaction. These decays, (12.5), are decays by the weak force. The weak force is so named because of the weakness (slowness) of the interaction. We have a picture of muon decay:

Pions are the least massive of the hadrons; they come in three varieties, $\{\pi^0, \pi^-, \pi^+\}$. Because pions are the least massive of hadrons, they cannot decay into other hadrons – no-where to go. The electrically neutral π^0 can decay electromagnetically into two photons $\pi^0 \rightarrow \gamma\gamma$, but, because of electric charge conservation, the electrically charged pions cannot decay into photons; they can decay by only the weak force.

It is entirely reasonably to expect that the muon might decay into an electron and a photon as $\mu^- \rightarrow e^-\gamma$, but it never does. This is part of the neutrino puzzle. The standard explanation given for the absence of $\mu^- \rightarrow e^-\gamma$ decays is that lepton number is conserved within each lepton generation so that, instead of having just one lepton number conservation law, we have three lepton number conservation laws, one for each generation of leptons.

Part of the neutrino puzzle is that such separate generation lepton number conservation laws seem not to be the case. Effectively, by a convoluted line of reasoning, the absence of the $\mu^- \rightarrow e^-\gamma$ decay mode indicates that neutrinos are massless. Other evidence says neutrinos have mass; so why do we not see the $\mu^- \rightarrow e^-\gamma$ decay mode?

The kaon, K^+, can decay in several weak decay ways. We have:

$$K^+ \rightarrow \mu^+ + \nu_\mu \qquad\qquad K^+ \rightarrow e^+ + \nu_e$$

$$K^+ \rightarrow \pi^0 + \mu^+ + \nu_\mu \qquad\qquad K^+ \rightarrow \pi^0 + e^+ + \nu_e$$

$$K^+ \rightarrow \pi^0 + \pi^+ \qquad\qquad K^+ \rightarrow \pi^+ + \pi^+ + \pi^-$$

$$K^+ \rightarrow \pi^0 + \pi^+ + \pi^0$$

$$(12.6)$$

The reader should notice that there are no neutrinos in three of the above decays, (12.6).

Violation of parity:

Parity is the name given by physicists to the idea that physics is the same if the physical apparatus is reflected in a mirror. The reader will recall that we spoke of physics being invariant under a rotation of the physical apparatus – a kettle boils at the same temperature with its spout pointing north as the temperature it boils at with its spout pointing west. It might seem reasonable to expect that a kettle would boil at the same temperature if it were reflected in a mirror. Well, it does boil at the same temperature if reflected in a mirror because boiling water is an electromagnetic phenomenon, but the reflected kettle would not boil at the same temperature if boiling water was a weak force phenomenon.

The decay of K^+ into two pions produces decay products of a different parity to the decay of K^+ into three pions. The weak force does not respect parity. To the weak force, right-handed is not a reflection of left-handed. We will see that the two quaternion algebras have this same violation of parity property in due course.

Prior to 1956, physicists believed that the universe would respect parity. It came as a great shock to discover that the weak force does not respect parity. The decay of the kaons played a major roll in the realisation of parity violation. We will have more to say of this later when we again tread our *"...road less travelled..."*.

The Standard model:

Within particle physics, there is an understanding called the Standard model. The Standard model is a lagrangian; that is, the Standard model is a mathematical expression which is a lagrangian rather like, but much longer than, the Dirac lagrangian above, (9.3). The Standard model is seen by physicists as a model of the universe. Physicists do not claim that the Standard model is the universe or even that it describes the universe; they simply claim that it models the quantum physics part of the universe.

The Standard model does not include classical physics; that is, the Standard model does not include gravity, the expanding universe, or Maxwell's classical electromagnetism. The Standard model does not explain why our space-time is 4-dimensional or why gravity exists or why we have one time dimension and three spatial dimensions. Indeed, there are many things which the Standard model does not include like why we have physical constants, why the universe had an inflationary beginning, or why there are three generations of fermions (for example: electron, muon, tau particle).

The Standard model has not been deduced from some great understanding of the universe or some great symmetry of mathematics or some principle of logic. The Standard model has been formulated by guessing. The guessing is highly educated guessing, and it is guided by respect for gauge symmetries and other most respectable principles, but, in the final analysis, the Standard model is just a guess at how particle physics works. There are nineteen numbers, parameters like the mass of the electron or the charge of the electron or the strength of the electromagnetic interaction (the fine structure constant) which have to be fed into the Standard model by hand to make it work. No-one knows why these parameters have the values they do, and no-one has found a way to deduce these values.

Having said all of the above, the Standard model is seen by physicists as a momumental triumph. Together with the assemblage of quantum field theory calculational techniques like Feynmann diagrams, the Standard model enables physicists to predict the behaviour of particles with outstanding accuracy – physicists predicted superconductivity and built the LHC, didn't they. There are aesthetic aspects of the Standard model which are very impressive, but, more importantly, the Standard model includes every observed particle physics phenomenon. Well, that is every particle physics phenomenon except one - massive neutrinos.

The Standard model, together with the electroweak part of that Standard model assumes that neutrinos are massless. Yet, it seems, there is evidence that neutrinos are not massless.

Neutrino mass:

From the 1930's until recently, the neutrino was believed to be massless. The reasoning was simple. We take the beta decay of a neutron and we measure the energy of the neutron going into the interaction. We then measure the energy of the electron coming out of the interaction. Since mass is energy, if the neutrino had mass, the energy of the outcoming electron would always be less than the energy of the incoming neutron by the amount of energy equivalent to the mass of the neutrino. Measurements of the energy of the outcoming electron show that, as accurately as can be measured, the outgoing electron sometimes has the same energy as the incoming neutron. Therefore, the neutrino must have zero mass. Such measurements are referred to as the 'high end energy spectrum of neutron decay'.

As accurately as we can measure, all types of neutrinos travel at the speed of light. This implies that neutrinos are massless.

Note: The speed of light is outside of our 4-dimensional space-time. Mass seems to be a phenomenon of our 4-dimensional space-tiime.

We do not know what the individual masses of the different types of neutrinos, $\{v_e, v_\mu, v_\tau\}$ are, and so perhaps the electron neutrino does have zero mass while the other two neutrinos have some mass. This is unlikely for it is very messy to claim that the three types of neutrinos are all the same type of particle when one is massless and the others are not massless.

Neutrino oscillations:

The nuclear processes within the sun are well understood. The nuclear processes within the sun produce electron neutrinos, and we are able to calculate just how many electron neutrinos from the sun will hit the Earth each second. However, when physicists counted the number of electron neutrinos hitting the Earth each second, they were perplexed by 'missing' electron neutrinos. Only one third of the expected electron neutrino flux from the sun actually hits the Earth. For years, physicists struggled to understand this inconsistency.

We now know, from many experiments, that the electron neutrinos seem to 'transform' into muon neutrinos and tau neutrinos as they travel from the sun to the Earth and that this is the reason for the deficit of electron neutrinos hitting the Earth.

Within our understanding of nuclear physics, we can calculate the probability of such a transition from, say, an electron neutrino into a muon neutrino. That probability is:

$$\text{Prob}\left(v_e \rightarrow v_\mu\right) = \sin^2\left(2\theta\right)\sin^2\left(\frac{\left(m_e^2 - m_\mu^2\right)L}{4E}\right) \qquad (12.7)$$

wherein L is the distance between where the neutrino is created as an electron neutrino and where the observer sits; this is often the distance between the sun which creates the electron neutrinos and the Earth where the observer sits. m_e is the mass of the electron neutrino, and m_μ is the mass of the muon neutrino. θ is not zero. E is the energy of the neutrino. Only the mass terms change for different neutrino transitions.

Looking at the above, (12.7), we see that, if the neutrinos were all massless, then there would be zero probability of an electron neutrino turning into a muon neutrino – great. Although often said, this is not quite correct; if the square of the neutrino field were massless there would be zero probability of an electron neutrino turning into a muon neutrino. Let us be accurate in our view.

The problem is that the observed deficit of electron neutrinos hitting the Earth inclines us to accept that electron neutrinos transform into muon neutrinos. Thus, the mass-squared difference cannot be zero implying the neutrinos cannot be massless. This habit of neutrinos transforming from one type of neutrino into another type of neutrino is known as neutrino oscillation.

Earth bound neutrino oscillation:
Observation of 'missing' neutrinos is not only with neutrinos from the sun. Neutrinos from energetic cosmic rays hitting the nitrogen and oxygen nucelii of the upper atmosphere are measured to be in the ratio of muon neutrinos to electron neutrinos of:

$$\nu_\mu : \nu_e = 2 : 1 \qquad (12.8)$$

for neutrinos descending to the Earth's surface, but, for neutrinos ascending from the Earth's surface, having first descended to the Earth's surface at the other side of the Earth and passed through the Earth, this ratio is:

$$\nu_\mu : \nu_e = 1 : 1 \qquad (12.9)$$

We have lost some muon neutrinos.

Similar effects are seen for neutrinos both produced and observed upon the Earth.

Lepton number and the Standard model:
The Standard model assumes that lepton number is conserved within each separate generation These three separate lepton number conservation laws are violated if neutrinos can convert into each other. Thus the Standard model assumes that neutrinos cannot oscillate into each other; this is the same as assuming that neutrinos have zero mass squared difference - see (12.7). Zero mass squared difference is taken to mean zero mass. If all neutrinos had zero mass or the same non-zero mass, they would not be able to oscillate into

each other. If all neutrinos had the same mass, then the three types of neutrino, $\{v_e, v_\mu, v_\tau\}$ would be indistinguishable.

Neutrino oscillation does not mean that lepton number conservation over the three generations is violated; it means only that lepton number conservation is violated within each separate generation.

It is possible to amend the Standard model in several ways which would permit the three neutrino types to have mass and thus to oscillate into each other. No-one knows which, if any, of these proposed amendments is the correct amendment.

Neutrino spin:

We recall that quaternion rotation, indeed rotation in any spinor space, is not rotation about an axis. We recall that according to our "*... road less travelled ...*" view, intrinsic spin is associated with angular momentum within a spinor space, quaternion space in the case of the electron. Having recalled all this, we now give the conventional view of neutrino spin.

Conventionally, it is assumed that the angular momentum associated with intrinsic spin is rotation about an axis. Because of the quantitisation of the direction of spin, this means that the axis of rotation will be either aligned with the direction of motion of the particle, the momentum vector of the particle, or opposite to the direction of motion of the particle. The alignment or opposite alignment is called the helicity of the particle defined as:

$$h_{\vec{p}} \equiv \frac{2\vec{J}\cdot\vec{p}}{|\vec{p}|} \tag{12.10}$$

wherein $\vec{J}$ is the classical angular momentum of the particle which is taken to be the same as the intrinsic spin of the particle and $\vec{p}$ is the 3-momentum of the particle. Technically, because we put the factor of 2 into the helicity definition, (12.10), the eigenvalues of the helicity operator are either $+1$ or -1 corresponding to the spin being either aligned or not aligned with the momentum vector. These eigenvalues would have been $\left\{-\dfrac{1}{2}, +\dfrac{1}{2}\right\}$ if we had not included the 2 in the definition. Thus the helicity of a particle is simply twice the value of the intrinsic spin in the direction of the particle's motion.

A particle with helicity eigenvalue +1 is often said to be right-handed while a particle with helicity eigenvalue −1 is often said to be left-handed.

There is a problem with using helicity (as a quantum number) to describe a particle. Imagine a person in a space-ship moving parallel to the particle. If the space-ship is moving faster than the particle, then the momentum vector of the particle is pointing backward with respect to the space-ship, but, if the space-ship is moving slower than the particle, then the momentum vector of the particle is pointing forward with respect to the space-ship. By slowing the velocity of the space-ship or by increasing the velocity of the space-ship, the person observing the particle can change the helicity of the particle. For this reason helicity is not a good parameter to use describing the particle.

However, if the particle is 'massless', which is taken to mean the same as 'moving at the speed of light', then no observer can move faster than the particle, and so, for massless particles, helicity would be a good parameter (quantum number) describing that particle.

We see here part of the neutrino problem. If neutrinos have mass, they cannot move at the speed of light and so helicity is not a good quantum number for a neutrino. None-the-less, it does seem that neutrinos have mass and move at the speed of light, in which case, in spite of them being massive, helicity would be a good quantum number for neutrinos.

Of course, if intrinsic spin is not rotation about an axis, then helicity as defined above, (12.10), is meaningless.

Chirality:

Within particle physics, there is the concept of chirality. Chirality is often presented as being a posh name for being right-handed or being left-handed, but that is not quite accurate. For a particle moving at the speed of light, chirality is taken to be the same as helicity which is the same as being right-handed or being left-handed. However, chirality is seen as a property that is intrinsic to the particle rather than dependent upon the particle's and observer's motion in our 4-dimensional space-time. For particles moving at less than the speed of light, helicity, like being right-handed or being left-handed, depends on the observer's velocity, but chirality, being intrinsic to the particle, remains the same regardless of the velocity of the observer. How does that work? We take the "*...road less travelled...*" view.

There are two quaternion algebras that derive from the $C_2 \times C_2$ finite group. We have the quaternions:

$$\mathbb{H} = \begin{bmatrix} a & b & c & d \\ -b & a & -d & c \\ -c & d & a & -b \\ -d & -c & b & a \end{bmatrix}$$

(12.11)

$$ij = k \qquad ji = -k$$
$$ik = -j \qquad ki = j$$
$$jk = i \qquad kj = -i$$

and the anti-quaternions:

$$\mathbb{H}_{Anti} = \begin{bmatrix} a & b & c & d \\ -b & a & d & -c \\ -c & -d & a & b \\ -d & c & -b & a \end{bmatrix}$$

(12.12)

$$ij = -k \qquad ji = k$$
$$ik = j \qquad ki = -j$$
$$jk = -i \qquad kj = i$$

Aside: At a simple level, we can see these two algebras as the two ways of fitting together three anti-symmetric variables into a 4×4 matrix such that the form of the matrix is closed under multiplication.

The thing to notice is that the commutation relations of the anti-quaternions are the reverse of the commutation relations of the quaternions. It is established mathematical mantra that the commutation relations of the quaternions are isomorphic as a Lie algebra to the Lie algebra $SU(2)$. It seems that we have two types of $SU(2)$ to match the two types of quaternions. We might refer to these as right-chiral-$SU(2)$ and left-chiral-$SU(2)$. Our *"...road less travelled..."* view will in due course identify these two types of quaternions with two types of electrons which are left-chiral electrons and right-chiral electrons. Such electron chirality does not exist in our 4-dimensional space-time, it is a phenomenon of quaternion space, and this is why we see only one type of electron in our 4-dimensional space-time. We will consider this in more detail when we look at electroweak theory.

Returning our attentions to the neutrino, we see that if a neutrino is assiociated with one of the two quaternion algebras, then a neutrino will be either left-chiral or right-chiral depending upon with which of the quaternion algebras it is associated.

Remarkably, all neutrinos, of all three types, are left-chiral; that is they have intrinsic spin anti-parrallel to their momentum vector. Similarly, all anti-neutrinos are right –chiral. It is believed that right-chiral neutrinos, and left-chiral anti-neutrinos, do not exist within our 4-dimensional space-time. We will see this result emerge from our *"...road less travelled..."* derivation of the Dirac equation in a later chapter.

Note: Quarks have mass. If quarks have mass, then, since they move at less than the velocity of light, they must exist if both left-handed and right-handed forms.

Of course, we remind the gentle reader that in our *"... road less travelled ..."* view, anything which moves at the speed of light is outside of our 4-dimensional space-time.

Anti-neutrinos:

In the Standard model, anti-neutrinos differ from neutrinos in two ways. One way is by their helicity; the other way is by their lepton number. As explained above, helicity is a questionable concept, and our *"... road less travelled ..."* view rejects it entirely. Helicity is meaningless for massive neutrinos. This means that massive anti-neutrinos differ from the corresponding massive neutrinos by only their lepton number.

If neutrinos oscillate into one another, then lepton number is not conserved within each separate generation, but might still be conserved overall. If overall lepton number is not conserved for massive neutrinos, then massive neutrinos are not distinguished from massive anti-neutrinos; if overall lepton number is not conserved neutrinos are their own anti-particles; if overall lepton number is conserved neutrinos are not their own anti-particles. We thus have two types of neutinos.

Neutrinos are fermions. Fermions are conventionally represented as a pair of complex numbers; anti-fermions are represented by the conjugate of the pair of complex numbers – see later. If the two complex numbers in the pair representing a particle are both real, with zero imaginary part, then that particle is its own anti-particle.

Neutrinos which are not their own anti-particles are called Dirac neutrinos; they correspond to solutions of the Dirac equation which are complex (two complex numbers) – see later chapter. Neutrinos which are their own anti-particles are called Majorana neutrinos after Ettore Majorana (1906-1959); they correspond to solutions of the Dirac equation which are real (two complex numbers with zero imaginary parts) – see later chapter. Ettore Majorana found solutions to the Dirac equation which are real.

The standard mantra is that no-one knows whether neutrinos are Majorana neutrinos with $v_i = \overline{v}_i$, or Dirac neutrinos with $v_i \neq \overline{v}_i$. This is the same as no-one knowing whether or not there exists a conserved lepton number.

Neutrinoless double beta decay:

In principle, there is a way to determine by experiment whether neutrinos are Dirac or Majorana. There are some atomic nuclei which cannot decay by ordinary single beta decay because the resultant nucleus is more massive than the initial nucleus. For example, the germanium nucleus Ge_{32}^{76} cannot decay by single beta decay into the arsenic nucleus AS_{33}^{76} because the arsenic nucleus is more massive than the germanium nucleus:

$$Ge_{32}^{76} \nrightarrow As_{33}^{76} + e^- + \overline{v}_e \qquad (12.13)$$

Other such nuclei are Mo_{42}^{100}, Te_{52}^{130} and Xe_{54}^{136}. However, because the nucleus of the element with two more protons than these elements is less massive than these nuclei, these nuclei can decay by beta decay if two neutrons become two protons simultaneously and the nucleus emits two electrons simultaneously:

$$Ge_{32}^{76} \rightarrow Se_{34}^{76} + 2e^- + 2\overline{v}_e \qquad (12.14)$$

This is called double beta decay. This kind of decay is very improbable, but it does happen; it has been observed in twelve different nuclides.[72]

According to the Standard model, the above decay, (12.14), is as shown on the forthcoming piece of most delightful art:

---

[72] See: Burgess & Moore The Standard model A Primer pg 419

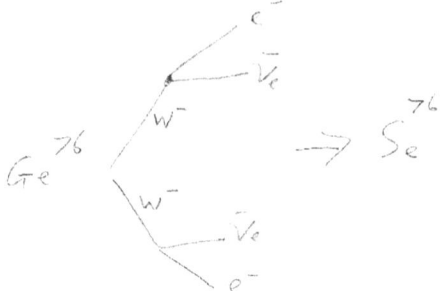

The above is called double beta decay.

A less likely decay mode, which has not been observed to date is the same as the above except that the two neutrinos mutually anihilate each other. This can happen only if the neutrino is its own anti-particle – if the neutrinos are Majorana, that is. This decay mode is called neutrinoless double beta decay; it is forbidden by the Standard model which assumes that neutrinos are of a Dirac nature. More fine art:

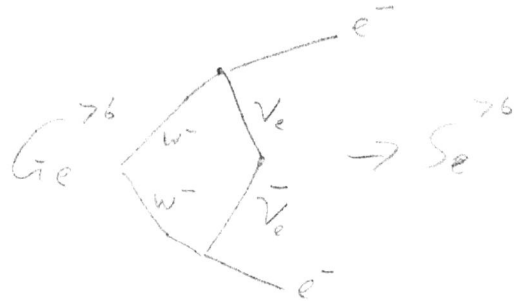

The observation of neutrinoless double beta decay would demonstrate the violation of lepton number conservation. We repeat that, to date, it has not been observed. An equally important point is that you do not get such lovely pictures in most physics books.

Summary of neutrinos:
Neutrinos are intimately associated with electrons. The reader might have associated the electron with only the electromagnetic interaction (force), but we have seen above that the electron also participates in

the weak interaction (force). When the electron participates in the weak interaction, it is asociated with an anti-electron neutrino.

Neutrinos are spin one half particles – they are fermions. Neutrinos have no electric charge. Neutrinos have lepton number, and they take part in the weak interaction.

Neutrinos seem to travel through our 4-dimensional space-time at the speed of light; this indicates that neutrinos have zero mass. There are other reasons like the absence of the $\mu^- \to e^- \gamma$ decay mode to think neutrinos are massless.

Neutrinos come in three different types corresponding to the three fermion generations. The 'missing' electron neutrinos from the sun, and other processes, indicate that neutrinos oscillate between the three different types; this oscillation indicates that neutrinos have mass.

If one type of neutrino can change into another type of neutrino, then the lepton number carried by neutrinos cannot be conserved within each generation of leptons. This means we have only one lepton number conservation law rather than a lepton number conservation law for each generation of leptons.

The standard mantra is that it is now generally accepted that neutrinos have mass; this might be an error.

The very successful Standard model of particle physics, together with the very successful electroweak theory, assumes that neutrinos have zero mass. This conflict between the Standard model and neutrino mass is a central conundrum of modern particle physics.

If lepton number is not conserved, then neutrinos are their own anti-particles; such neutrinos are called Majorana neutrinos. If lepton number is conserved, then neutrinos are not their own anti-particles; such neutrinos are called Dirac neutrinos. The Standard model assumes that neutrinos are Dirac neutrinos.

Chapter 13

The Electron Equations

Within quantum mechanics and quantum field theory, the electron, and all spin one half particles, are conventionally taken to be represented by an ordered pair of complex numbers presented as a vector:

$$Electron = \begin{bmatrix} a+ib \\ c+id \end{bmatrix} \qquad (13.1)$$

This fact is sometimes lost to students because the Schrödinger equation 'sloppily' takes the electron to be 'spinless' and to be represented by a single complex number, and students are introduced to quantum mechanics through the Schrödinger equation.

Why is the electron represented by a pair of complex numbers? Why not a single complex number or perhaps a 3-dimensional complex number[73] or a 5-dimensional complex number or just an ordered set four real numbers or a trio of complex numbers? If I were a mathematician and if I were to choose a mathematical object to represent an electron, I doubt that I would choose a pair of complex numbers.

With the choice of a pair of complex numbers, there comes the concept of a complex Hilbert space. This is like a normal 2-dimensional space except that it has complex axes rather than real axes. The states of the electron correspond to points, ordered pairs of complex numbers, in the complex Hilbert space. Your author opines that there is no such thing as a complex axis, and that complex Hilbert

---

[73] See: Dennis Morris Complex Numbers The Higher Dimensional Forms (2nd edition).

spaces ought not to be part of a physical theory. Still, your author could be wrong. Everyone else seems to accept complex axes.

Quaternion perhaps:

We know that a quaternion can be written as an ordered pair of complex numbers, and so it seems that the electron might be represented by a quaternion and that the ordered pair of complex numbers is just obscure notation for a quaternion. This view replaces the complex Hilbert space with a quaternion space in which each state of the electron is a point in quaternion space. Yet still we have the question: Why a quaternion?

The energy momentum relation of special relativity is:

$$E^2 = p^2 + m^2$$
$$p^2 = \vec{p} \cdot \vec{p} = p_x p_x + p_y p_y + p_z p_z$$

$$(13.2)$$

We will see in a later chapter that the square root of $\left(p^2 + m^2\right)$ is a quaternion, and that this leads directly to the relativistic electron equation, the Dirac equation. And so, directly out of special relativity, we have a quaternion, which can be written as a pair of complex numbers. It seems that this might be why electrons are represented by an ordered pair of complex numbers.

The electron equations of quantum physics:

There are three equations within modern physics which describe the electron. These equations are the Schrödinger equation, the Pauli-Schrödinger equation, and the Dirac equation. These equations are all different manifestations of the same equation. There is also the Klein-Gorden equation which describes zero spin particles.

The starting point of all three electron equations, and of the Klein-Gorden equation, is the energy momentum relation. The energy momentum relation has a classical form and a relativistic form. These are respectively the classical energy momentum relation:

$$E = \frac{p^2}{2m} + V \qquad (13.3)$$

and the relativistic energy momentum relation:

$$E^2 = p^2 + m^2 \qquad (13.4)$$

The Schrödinger equation and the Pauli-Schrödinger equation are both non-relativistic equations meaning that they are based upon the classical energy momentum relation, (13.3); they describe electrons moving at slow velocities. The Dirac equation, and the Klein-Gorden equation, are relativistic equations meaning that they are based upon the relativistic energy momentum relation, (13.4).

Clearly, the relativistic equations, the Dirac equation and the Klein-Gorden equation are the correct forms of the equations of particle physics. The non-relativistic equations are used because they are simpler and because they are 'accurate enough' for common usage. The Schrödinger equation is used as a padagogic tool to introduce students to quantum physics without all the complexities of intrinsic spin, largangians, and gauge theory.

The Schrödinger equation:
We begin with the non-relativistic relation:

$$E = \frac{p^2}{2m} + V \qquad (13.5)$$

wherein E is the energy, p is the momentum, m is the mass of the particle (electron mass in our case) and V is the potential which may be a function of position and time. We substitute into this equation the two quantum mechanical operators for energy and momentum:

$$E = i\hbar \frac{\partial}{\partial t} \qquad \& \qquad p = -i\hbar \frac{\partial}{\partial x} \qquad (13.6)$$

and we give these operators a complex field (a complex function of space and time), $\psi(t,x) \in \mathbb{C}$, to operate upon. The Schrödinger equation is an equation in which the variable, complex field, $\psi(t,x)$ is a single complex number

$$\psi_S(t,x) = a(t,x) + ib(t,x) \tag{13.7}$$

which varies from point to point in our 4-dimensional space-time.

This gives:

$$i\hbar \frac{\partial}{\partial t}\psi(t,x) = -\frac{\hbar^2}{2m}\frac{\partial^2}{\partial x^2}\psi(t,x) + V(t,x)\psi(t,x) \tag{13.8}$$

This is the time dependent Schrödinger equation, TDSE. It is time dependent because the function, $\psi(t,x)$, and the potential, $V(t,x)$, depend on time. The field, function, $\psi(t,x) \in \mathbb{C}$ is called the wave function. The TDSE is a linear differential equation, and so its different solutions can be added to form other solutions.

Relativity is based on the understanding that space and time are equivalent. We see in the Schrödinger equation, (13.8), that we have the first differential of time but we have the second differential of space. Because time and space are treated unequally in the Schrödinger equation, the Schrödinger equation cannot be concordent with relativity theory. Thus, we know, as Schrödinger knew at the time he wrote this equation, that the Schrödinger equation is wrong. We will see in the next chapter that Dirac corrected this problem.

So, all we have done is taken the classical definition of kinetic energy and potential energy, (13.5), and put quantum mechanical operators in place of the variables for energy and momentum; this is the TDSE.

Perhaps confusingly, if the potential in (13.8) is independent of time, the time dependent Schrödinger equation is:

$$i\hbar \frac{\partial}{\partial t}\psi(t,x) = -\frac{\hbar^2}{2m}\frac{\partial^2}{\partial x^2}\psi(t,x) + V(x)\psi(t,x) \tag{13.9}$$

This is also called the time dependent Schrödinger equation, TDSE. For example, the space around an atomic nucleus has an electromagnetic potential that is independent of time. Now, if the potential is independent of time, as in (13.9), then the TDSE has solutions of the form:

$$\psi(t,x) = \Phi(x)e^{-i\frac{Et}{\hbar}} \tag{13.10}$$

To get solutions of the TDSE, we just need to find $\Phi(x)$ and multiply these expressions by $e^{-i\frac{Et}{\hbar}}$. Now, $\Phi(x)$ satisfies the equation:

$$-\frac{\hbar^2}{2m}\frac{\partial^2}{\partial x^2}\Phi(x) + V(x)\Phi(x) = E\Phi(x) \tag{13.11}$$

This equation, (13.11), is called the time independent Schrödinger equation, TISE. E is just a real number which is an amount of energy.

The probability density of a solution of the form (13.10) is:

$$|\psi(t,x)|^2 = \psi^*(t,x)\psi(t,x)$$
$$= \Phi^*(x)e^{i\frac{Et}{\hbar}}\Phi(x)e^{-i\frac{Et}{\hbar}} \tag{13.12}$$
$$= \Phi^*(x)\Phi(x)$$

We see that the probability density is independent of time. Thus, solutions of the form (13.10) are unchanging in time; they are called stationary solutions of the TDSE – stationary is unchanging in time.

The E in (13.11) is just a real number; it is the energy of the particle (electron) which is in the stationary state. Basically, E is the energy of the electron in orbit about an atomic nucleus. There are different solutions to the TISE, (13.11), these different solutions correspond to different values of E; these are the different energy levels of electrons in orbit around an atomic nucleus, or in any electromagnetic potential which is independent of time.

The TDSE is used to discover the energy levels of a particle of mass m in a potential V. We insert the given potential into the TDSE and solve the TISE to get the energy levels.

Often, solutions of the TDSE are of the form of waves like $\psi(t,x) = A\sin(kx)e^{i\frac{Et}{\hbar}}$. This is why $\psi(t,x)$ is called a wave function. Looking at the basic form of (13.11), which is:

$$\frac{\partial^2}{\partial x^2}\Phi(x) = -E\frac{2m}{\hbar^2}\Phi(x) \qquad (13.13)$$

and bearing in mind:

$$\frac{\partial^2}{\partial x^2}(\sin kx + \cos kx) = -k^2(\sin kx + \cos kx) \qquad (13.14)$$

We see that it is no surprise to find solutions which are wave functions.

The electron as described by the TDSE:
In the view of the TDSE, an electron is a superposition of stationary waves represented as stationary wave functions. The reader will recall that the TDSE is a linear differential equation; as such, if its solutions are waves, then any sum of those waves will also be a solution of the TDSE. Such a sum of waves will form, if we are careful in our choice of waves and coeficients, a wave packet. Such a superposition is basically a sum of different wave functions in which each wave function is multiplied by a complex coeficient. We have:

$$\Psi = e^{-i\frac{Et}{\hbar}}(c_1\psi_1 + c_2\psi_2 + c_3\psi_3 + ...) \qquad (13.15)$$

The coeficients are chosen to form a wave packet. We see that the TDSE takes the view that electrons are basically waves and not particles but that they gain a particle like structure as a wave packet through carefully chosen coeficients. This is nothing like the view of an electron as a point particle that scattering experiments have presented to physicists.

If we have zero potential, that is a freely moving electron, then a solution of the TDSE is the de Broglie wave equation (we assume the classical definition of kinetic energy):

$$\psi = Ae^{-i\left(\frac{Et-px}{h}\right)} \equiv \begin{bmatrix} A & 0 \\ 0 & A \end{bmatrix} \begin{bmatrix} \cos\left(\dfrac{Et-px}{h}\right) & \sin\left(\dfrac{Et-px}{h}\right) \\ -\sin\left(\dfrac{Et-px}{h}\right) & \cos\left(\dfrac{Et-px}{h}\right) \end{bmatrix}$$

(13.16)

We might have lost a minus sign when we took the equivalence, but it is only the direction of rotation. Looking at (13.16), we see that an electron is something to do with rotation in the complex plane. The industrious reader is reminded that, according to Noether's theorem, electric charge is due to invariance under rotation in the complex plane, $U(1)$. Rotation in the complex plane is closely connected to waves.

When all is said and done, the TDSE is a wave equation of the form:

$$f = \frac{1}{2m}\frac{\hbar}{\lambda^2}$$

(13.17)

wherein f is frequency and λ is the wavelength. This is just a relation between frequency, which is the time dependence of the wave, and wavelength, which is the spatial dependence of the wave.

Since the TDSE is a wave equation, its solutions will be waves. If the solution of the TDSE describes an electron, then that electron is a wave or a linear sum of solutions (waves) – a wave packet. It is the TDSE that has led to the view that an electron is a wave/wave packet.

The wave view of the electron leads directly to the different energy levels of electron orbits around atomic nuclei. The orbits of electrons around atomic nuclei are such that the length of the orbit is a complete number of wavelenths. It is hard to see how a point particle could fit into only orbits of integer wavelength circumference.

The solutions of the TDSE which are the electron orbits are stationary solutions which are solutions of the TISE – that's time independent. There is no time variable in the TISE. It seems that we have wave like electrons associated with the absence of the time variable. Of course, the quaternion distance function coincides with the distance function of our 4-dimensional space-time if the time variables can be set to zero.

The Schrödinger equation successfully describes all non-relativistic atomic phenomena except those involving magnetism. To include magnetism, we need to use the Pauli-Schrödinger equation.

The Pauli Schrödinger equation:

The Pauli-Schrödinger equation is a generalisation of the Schrödinger equation introduced by Pauli to include the electron's spin within the equation. Whereas the Schrödinger equation describes 'spinless' electrons, the Pauli-Schrödinger equation describes electrons with spin. The variable, field, within the Pauli-Schrödinger equation is a pair of complex numbers which vary independently with the position in our 4-dimensional space-time:

$$\psi_{PS}(t,x) = \begin{bmatrix} a(t,x) + ib(t,x) \\ c(t,x) + id(t,x) \end{bmatrix} \tag{13.18}$$

Such a pair of complex numbers is called a spinor by physicists[74]. The Pauli-Schrödinger equation therefore has a spinor field as its variable. Of course, the reader is aware of the fact that a quaternion can be written as a pair of complex numbers, and it ought not to surprise the reader that your author takes this pair of complex numbers to be a quaternion written in poor notation.

---

[74] Your author is of the view that a spinor is a type of complex number, and so this pair of complex numbers at first appears to not be a spinor. It can be viewed as a spinor, a quaternion, but written in misleading notation.

In an electromagnetic field $\{\vec{E}, \vec{B}\}$ with potentials $\{V, \vec{A}\}$, the Schrödinger equation is[75]:

$$i\hbar \frac{\partial}{\partial t} \psi = \frac{1}{2m} \left(-i\hbar \frac{\partial}{\partial x^\mu} - e\vec{A} \right)^2 \psi - eV\psi \qquad (13.19)$$

We replace $\left(-i\hbar \frac{\partial}{\partial x^\mu} - e\vec{A} \right)^2$ by $\left(\pi^2 - \hbar e(\vec{\sigma} \cdot \vec{B}) \right)$ to get the Pauli-Schrödinger equation which is:

$$i\hbar \frac{\partial}{\partial t} \psi = \frac{1}{2m} \left(\pi^2 - \hbar e(\vec{\sigma} \cdot \vec{B}) \right) \psi - eV\psi \qquad (13.20)$$

Within this equation, the term $(\vec{\sigma} \cdot \vec{B})$ is a 2×2 matrix:

$$(\vec{\sigma} \cdot \vec{B}) = \begin{bmatrix} 0 & B_x \\ B_x & 0 \end{bmatrix} + \begin{bmatrix} 0 & -iB_y \\ iB_y & 0 \end{bmatrix} + \begin{bmatrix} B_z & 0 \\ 0 & -B_z \end{bmatrix} \qquad (13.21)$$

And:

$$\pi^2 = p^2 + e^2 A^2 - e\left(\vec{p} \cdot \vec{A} + \vec{A} \cdot \vec{p} \right) \qquad (13.22)$$

Because $(\vec{\sigma} \cdot \vec{B})$ is a 2×2 matrix, it must act on a 2-component vector of the form of (13.18). In fact, the presence of the imaginary numbers within the 2×2 matrix, (13.21), betrays the fact that this is really a 4×4 matrix that would act on a real 4-component vector. It is the established habit of writing a complex number as $a + ib$ rather than as a 2×2 matrix which leads to this confusion over notation.

The Pauli-Schrödinger equation is still a linear wave equation like the Schrödinger equation, and so it still sees electrons as waves or as wave packets. The Pauli-Schrödinger equation differs from the Schrödinger equation because it includes the spin of the electron which is absent in the Schrödinger equation; it does this by having a

---

[75] See: Pertti Lounesto Clifford Algebras and Spinors pg 51

variable which is two complex numbers rather than the single complex number which the Schrödinger equation uses.

The Klein-Gorden equation:

Above, we derived the Schrödinger equation from the non-relativistic energy momentum relation $E = \dfrac{p^2}{2m}$. Suppose we do the same with the relativistic energy momentum relation:

$$E^2 = p^2 + m^2 \qquad (13.23)$$

wherein we have set the speed of light to unity. Just as we did to form the Schrödinger equation, we substitute into this equation, (13.23), the two quantum mechanical operators for energy and momentum:

$$E = i\hbar \frac{\partial}{\partial t} \qquad \& \qquad p = -i\hbar \frac{\partial}{\partial x} \qquad (13.24)$$

and we give these operators a function, complex field, $\phi(t,x) \in \mathbb{C}$, to operate upon. This gives:

$$-\frac{\partial^2}{\partial t^2}\phi + \frac{\partial^2}{\partial x^2}\phi = m^2\phi \qquad (13.25)$$

This is the Klein-Gorden equation. The Klein-Gorden equation does not describe the electron. The Klein-Gorden equation describes spin zero particles. We include the Klein-Gorden equation here to show its relationship to the Schrödinger equation and other equations that do describe the electron.

The Klein-Gorden equation is an equation in which the variable, the field, is a single complex number. Confusingly, this is called a scalar field even though a complex number is, in your author's view, a spinor[76]. By this nomenclature, the Schrödinger equation would be

---

[76] Other mathematicians cannot decide whether a single complex number is a spinor or a scalar. There is much ambivalence among Clifford algebraists in this regard.

seen as having a scalar field as its variable, and yet an electron is unanimously agreed to be a spinor field. Of course, the Schrödinger equation describes a spin less electron which, in a sense, is a spin zero particle.

The important point is that the Klein-Gorden equation is consistent with the relativistic energy momentum relation, (13.23).

The Dirac equation:

It is often said that the Dirac equation is the square root of the Klein-Gorden equation. By this it is meant that, by a few mathematical manipulations[77], the Klein-Gorden equation can be derived from the Dirac equation as the 'kind of' square of the Dirac equation. The significance of being able to derive the Klein-Gorden equation from the Dirac equation is that this implies that the Dirac equation is consistent with the relativistic energy momentum relation (13.23).

Thus, we can say the Dirac equation is based on the relativistic energy momentum relation:

$$E^2 = p^2 + m^2 \qquad (13.26)$$

This often-proclaimed statement is not quite true. The Dirac equation is based upon the relativistic energy momentum relation:

$$E = \pm(p + m) \qquad (13.27)$$

Straight away, we see that we have negative energies as well as positive energies. Although the reader will not believe the author at this point, (13.27) is a square root of (13.26). We give the details in the following chapters.

The Dirac equation is an equation which describes relativistic fermions; that is electrons or neutrinos or quarks or their equivalents in the two other generations.

---

[77] See: David McMahon Quantum Field Theory DeMystified pg 88

The variable in the Dirac equation is called a Dirac spinor[78], and this variable is taken to represent the electron/positron field. As with the Pauli-Shrödinger equation, the electron is represented by a pair of complex numbers; the positron is also represented by a pair of complex numbers. All four complex numbers are put together to form the Dirac spinor. We have the Dirac spinor:

$$\Psi_{Dirac} = \begin{bmatrix} a+ib \\ c+id \\ e+if \\ g+ih \end{bmatrix} \equiv \begin{bmatrix} \begin{bmatrix} a+ib \\ c+id \end{bmatrix} \\ \begin{bmatrix} e+if \\ g+ih \end{bmatrix} \end{bmatrix} \equiv \begin{bmatrix} e^- \\ e^+ \end{bmatrix} \tag{13.28}$$

Your author considers a spinor to be a type of n-dimensional complex number with one real axis and $(n-1)$ imaginary axes (like a quaternion or a eucldean complex number). In spite of its name, a Dirac spinor is not what your author considers to be a spinor; your author takes the view that the Dirac spinor is a pair of spinors, a pair of quaternions, collated together. Such a pair of quaternions is not a division algebra, and so the Dirac spinor is a mathematical object that does not really exist. Since the Dirac equation is very successful and is central to QFT, this is worrying to folk who worry about such things.

The components of the Dirac spinor vary with the frame of reference. They must do this if the Dirac spinor is to be invariant under change of the velocity of the observer – special relativity.

The Dirac equation is usually presented as being the equations of motion, the Euler-Lagrange equations, derived from the Dirac lagrangian. The Dirac lagrangian[79] is:

$$\mathcal{L}_{Dirac} = \overline{\Psi}\left(i\gamma^\mu \partial_\mu - m\right)\Psi$$
$$\overline{\Psi} = \Psi^\dagger \gamma^0 \tag{13.29}$$

---

[78] This is a different usage of the word spinor to that used by your author.
[79] The following few sentences are put nicely in 'Dirac, Majorana and Weyl fermions by Palash B. Pal arXiv.1006.1718v2 – Oct 2010

and the Dirac equation derived from this lagrangian is:

$$\left(i\gamma^{\mu}\partial_{\mu}-m\right)\Psi=0 \qquad (13.30)$$

Written less concisely, this is:

$$i\hbar\left(\gamma^0\frac{1}{c}\frac{\partial\psi}{\partial t}+\gamma^1\frac{\partial\psi}{\partial x}+\gamma^2\frac{\partial\psi}{\partial y}+\gamma^3\frac{\partial\psi}{\partial z}\right)=mc\psi \qquad (13.31)$$

We will explain what the funny symbols are shortly.

We do not have to rely upon the Dirac lagrangian to get the Dirac equation. The Dirac equation is just the Schrödinger equation:

$$i\frac{\partial}{\partial t}\Psi=H\Psi \qquad (13.32)$$

with:

$$H=\gamma^0\left(\gamma^i p_i+m\right) \qquad (13.33)$$

and so we see a connection between these two equations. The lagrangian formulation is most commonly used. Having said all this, perhaps the best way to view the Dirac equation is as the square root of the Klein Gorden equation.

In the non-relativistic limit, the Dirac equation for an electron in an electromagnetic field $A^{\mu}=\left(A^0,\vec{A}\right)$ reduces to the Pauli-Schrödinger equation[80].

Having shown the connections between the three equations which describe the electron, the Schrödinger equation, the Pauli-Schrödinger equation and the Dirac equation, we must now explain the funny symbols used in the Dirac equation.

---

[80] Halzen & Martin Quarks and Leptons ISBN: 0-471-88741-2 page 107.

The Einstein notation:

The sumbol ∂_x is a shorthand way of writing differentiate with respect to x. The symbol $\gamma^\mu \partial_\mu$ means summation over the range of values of the indices, μ. In the Dirac equation, the indices are the variables $\{t, x, y, z\}$. We have:

$$\gamma^\mu \partial_\mu = \gamma^0 \frac{\partial}{\partial t} + \gamma^1 \frac{\partial}{\partial x} + \gamma^2 \frac{\partial}{\partial y} + \gamma^3 \frac{\partial}{\partial z} \qquad (13.34)$$

Of course, the differentials act upon the Dirac spinor field, Ψ.

The Dirac adjoint:

The Dirac adjoint spinor is written as $\overline{\Psi}$. This is a kind of conjugate like a quaternion conjugate to a quaternion or a complex number conjugate to a complex number. We derive this in more detail in the next chapter. Looking at the Dirac spinor as four pairs of complex numbers, (13.28), we might expect the 'conjugate' of this spinor to be a little different from a normal conjugate. With the Dirac spinor as:

$$\Psi = \begin{bmatrix} a+ib \\ c+id \\ e+if \\ g+ih \end{bmatrix} \qquad (13.35)$$

we have:

$$\Psi^\dagger = \begin{bmatrix} a-ib & c-id & e-if & g-ih \end{bmatrix}$$
$$\overline{\Psi} = \Psi^\dagger \gamma^0 = \begin{bmatrix} a-ib & c-id & -e+if & -g+ih \end{bmatrix} \qquad (13.36)$$

Within this, we have:

$$\gamma^0 = \begin{bmatrix} 1 & 0 & 0 & 0 \\ 0 & 1 & 0 & 0 \\ 0 & 0 & -1 & 0 \\ 0 & 0 & 0 & -1 \end{bmatrix} \tag{13.37}$$

As we will see in a later chapter, this 'weird' form of adjoint is a consequence of writing a quaternion as a pair of complex numbers.

The Dirac gamma matrices:

The Dirac matrices, also called the gamma matrices or the Dirac gamma matrices, are 4×4 matrices[81]. The essence of the gamma matrices is the two conditions:

$$\gamma^0\gamma^0 = +I, \qquad \gamma^1\gamma^1 = -I, \qquad \gamma^2\gamma^2 = -I, \qquad \gamma^3\gamma^3 = -I$$
$$\gamma^i\gamma^j = -\gamma^j\gamma^i \qquad for \qquad i \neq j$$
$$\tag{13.38}$$

wherein I is the identity matrix.

We also have:

$$\gamma_0{}^\dagger = \gamma_0, \qquad \gamma_1{}^\dagger = -\gamma_1, \qquad \gamma_2{}^\dagger = -\gamma_2, \qquad \gamma_3{}^\dagger = -\gamma_3$$
$$\tag{13.39}$$

The position of the indices on the gamma matrices, superscript or subscript, is meaningless, and the placement of the indices is for convenience of presentation only. The first of these conditions, (13.38), is that γ^0 is the square root of plus one, and the other three gamma matrices are square roots of minus one. This corresponds exactly to a unit quaternion in which the real variable is the square root of plus one and the three imaginary variables are square roots of minus one. The second of these conditions is that the conjugate of the γ^0 matrix is itself (γ^0 is real) and that the conjugates of the other

---

[81] There is another different set of matrices which are also known as the Dirac matrices – see later chapter.

three gamma matrices is the negative of the gamma matrix. This is also a property of the quaternion variables.

The first condition upon the gamma matrices, (13.38), is necessary if the Dirac equation is to comply with the relativistic energy momentum relation $E^2 = p^2 + m^2$ - see next chapter. The second condition, (13.39), leads to a Hamiltonian (expression for energy) which has real values of energy (rather than complex values); to be technical, the second condition implies that the Dirac Hamiltonian is hermitian.

There are an infinite number of sets of 4×4 matrices which satisfy the above two conditions, (13.38) & (13.39). It is a mathematical theorem that if there are two sets of gamma matrices such that both sets satisfy the conditions (13.38) & (13.39), then these two sets will be related by a similarity transformation involving a unitary matrix. A standard set of gamma matrices often found in text books is:

$$\gamma^1 = \begin{bmatrix} 0 & 0 & 0 & 1 \\ 0 & 0 & 1 & 0 \\ 0 & -1 & 0 & 0 \\ -1 & 0 & 0 & 0 \end{bmatrix} \qquad \gamma^2 = \begin{bmatrix} 0 & 0 & 0 & -i \\ 0 & 0 & i & 0 \\ 0 & i & 0 & 0 \\ -i & 0 & 0 & 0 \end{bmatrix} \qquad (13.40)$$

$$\gamma^3 = \begin{bmatrix} 0 & 0 & 1 & 0 \\ 0 & 0 & 0 & -1 \\ -1 & 0 & 0 & 0 \\ 0 & 1 & 0 & 0 \end{bmatrix} \qquad \gamma^0 = \begin{bmatrix} 1 & 0 & 0 & 0 \\ 0 & 1 & 0 & 0 \\ 0 & 0 & -1 & 0 \\ 0 & 0 & 0 & -1 \end{bmatrix} \qquad (13.41)$$

The reader is free to use any set of four 4×4 matrices which satisfy the above two conditions, (13.38) & (13.39). One can obtain such matrices by the similarity transformation $W^{-1}\gamma^i W$ where W is any 4×4 matrix with an inverse. Okay, I'll 'fess up; we will be using the quaternion matrices.

If we were to choose the gamma matrices to be the quaternion matrices, then the gamma matrices would have been chosen to be the basis units of a division algebra. Not all choices of gamma matrices have this 'basis units of a division algebra' property. Take the four gamma matrices above, (13.40) & (13.41), multiply each gamma matrix by a real variable and add them to get:

$$\begin{bmatrix} a & 0 & c & d-ib \\ 0 & a & d+ib & -c \\ -c & -d+ib & -a & 0 \\ -d-ib & c & 0 & -a \end{bmatrix} \qquad (13.42)$$

Multiplying this matrix, (13.42), by itself gives:

$$\begin{bmatrix} a^2-b^2-c^2-d^2 & 0 & 0 & 0 \\ 0 & a^2-b^2-c^2-d^2 & 0 & 0 \\ 0 & 0 & a^2-b^2-c^2-d^2 & 0 \\ 0 & 0 & 0 & a^2-b^2-c^2-d^2 \end{bmatrix}$$
$$(13.43)$$

This simply demonstrates that the four gamma matrices we used satisfy the conditions required to be a gamma matrix. We note that this matrix, (13.43), is exactly what we would get if we squared a quaternion written with the same variables.

However, if we form another matrix like (13.42) but using the variables $\{e, f, g, h\}$ in place of the variables $\{a, b, c, d\}$ and we take a product of these two matrices, Prod, we get the elements of Prod as, for example:

$$\mathrm{Prod}_{[1,1]} = ae - bf - cg - dh + i(bh - df)$$
$$\mathrm{Prod}_{[4,4]} = ae - bf - cg - dh + i(-bh + df) \qquad (13.44)$$
$$\mathrm{Prod}_{[1,2]} = dg - ch + i(cf - bg)$$

For multiplicative closure, we would need:

$$\mathrm{Prod}_{[1,1]} = -\mathrm{Prod}_{[4,4]}$$
$$\mathrm{Prod}_{[1,2]} = 0 \qquad (13.45)$$

The elements of Prod given above, (13.44), do not satisfy the requirement of multiplicative closure.

This set of gamma matrices, (13.40) & (13.41), are not multiplicatively closed when used to form a single matrix. The set of gamma matrices corresponding to the quaternion unit matrices does have the property of multiplicative closure, and all other attributes of a division algebra. We see that not all sets of gamma matrices are equal. Of course, all sets of gamma matrices will satisfy the conditions required to be gamma matrices, but we could add the 'form a division algebra' condition if we wanted; this would leave us only the quaternion unit matrices and the anti-quaternion unit matrices.

Majorana gamma matrices:

There is a representation of the gamma matrices in which all the gamma matrices are imaginary. This representation was first found by Majorana; it is:

$$\gamma^0 = \begin{bmatrix} 0 & 0 & 0 & -i \\ 0 & 0 & i & 0 \\ 0 & -i & 0 & 0 \\ i & 0 & 0 & 0 \end{bmatrix} \qquad \gamma^1 = \begin{bmatrix} 0 & i & 0 & 0 \\ i & 0 & 0 & 0 \\ 0 & 0 & 0 & i \\ 0 & 0 & i & 0 \end{bmatrix}$$

$$\gamma^2 = \begin{bmatrix} 0 & 0 & 0 & -i \\ 0 & 0 & i & 0 \\ 0 & i & 0 & 0 \\ -i & 0 & 0 & 0 \end{bmatrix} \qquad \gamma^3 = \begin{bmatrix} i & 0 & 0 & 0 \\ 0 & i & 0 & 0 \\ 0 & 0 & -i & 0 \\ 0 & 0 & 0 & -i \end{bmatrix}$$

(13.46)

The significance of this Majorana representation, (13.46), of the gamma matrices is that it makes the Dirac equation, (13.30), into a real equation with real solutions; there is an i in the Dirac equation which multiplies each gamma matrix. The real solutions are called Majorana fermions.

The anti-particle of a fermion is taken to be the conjugates of the pair of complex numbers representing the fermion. Since the Majorana solutions are real, they are equal to their own conjugates and so a Majorana fermion is its own anti-particle. Since we have observed the electron's anti-particle the positron, we know that electrons are not Majorana fermions, but neutrinos might be Majorana fermions.

The Dirac equation leads to the Klein-Gorden equation:

It is the first condition upon the gamma matrices, (13.38), that guarantees that the Klein-Gorden equation can be derived from the Dirac equation and thus this condition also guarantees that the Dirac equation is consistent with the relativistic energy momentum relation, (13.23).

Chapter 14

The Pauli Exclusion Principle

Within quantum theory, all particles must have an amount of intrinsic spin which is a multiple of $\frac{1}{2}$. Particles with intrinsic spin which is an odd multiple of $\frac{1}{2}$ are called fermions and particles with intrinsic spin which is an even multiple of $\frac{1}{2}$ are called bosons.

The Pauli exclusion principle is the statement that two fermions, spin one half particles like the electron the neutrino and quarks, cannot simultaneously occupy the same quantum state. For electrons, this means that it is not possible for two electrons in an atom to have the same values of the all the four quantum numbers which are:

a) The principle quantum number, n.
b) The angular momentum quantum number, l.
c) The magnetic quantum number, m_l.
d) The spin quantum number, m_s.

There is a nomenclature confusion in that fermions are any particle whose spin is not an interger. For example, particles with spin:

$$\left\{ ...-\frac{5}{2}, -\frac{3}{2}, -\frac{1}{2}, \frac{1}{2}, \frac{3}{2}, \frac{5}{2}, ... \right\} \tag{14.1}$$

are all fermions. Similarly, any particle whose spin is a whole number:

$$\{ ...-2, -1, 0, 1, 2, ... \} \tag{14.2}$$

is a boson.

It is often the case that electrons within an atom have the same principle quantum numbers, the same angular momentum quantum numbers, and the same magnetic quantum numbers but have different spin quantum numbers. There are only two spin states for an electron; these are spin up, $+\frac{1}{2}$, and spin down, $-\frac{1}{2}$. This means that we have pairs of electrons sharing the same principle quantum numbers, angular momentum quantum numbers, and magnetic quantum numbers within an atom. This is known as electron pairing within an atom.

Bosons, spin one particles like the photon and the $W^{\pm}, Z^{0}$ particles, are not affected by the Pauli exclusion principle. Any number of bosons can simultaneously occupy the same quantum state.

Because they are composed of three spin one half quarks, protons and neutrons are fermions, and they are subject to the Pauli exclusion principle. Mesons, which are composed of two spin one half quarks, are bosons. Some atoms are fermions; for example, helium-3 has spin one half and is therefore a fermion and subject to the Pauli exclusion principle. Other atoms are bosons; for example helium-4 has spin zero and is therefore a boson and not subject to the Pauli exclusion principle. Within a super-conductor, electrons forn Cooper pairs; such a pair of electrons is a boson.

The Pauli exclusion principle was formulated by Wolfgang Pauli (1900-1958) in 1925.

The solidity of matter:
It is because of the Pauli exclusion principle that electrons are forced to 'stack up' in an atom rather than all fall together into the lowest energy state. This 'stacking up' of electrons is what we percieve as the solidity of matter. This is why two solid objects cannot occupy the same space at the same time even though both objects are entirely empty space. This understanding of the solidity of matter was first voiced by Paul Ehrenfest (1880-1933) in 1931. The mathematical

formulation of Ehrenfest's understanding was given by Freeman Dyson (1923-) and Andrew Lenard in 1967.[82]

It is the Pauli exclusion principle which leads to the stability of neutron stars. This was shown by Elliot Lieb (1932-) et al in 1995.[83] Within neutron stars, it is the neutrons which, being fermions, must be 'stacked up' rather than the electrons of ordinary matter.

Given the Pauli exclusion principle, one might expect that matter could never be 'crushed' completely, but according to general relativity, and seemingly in accordance with astronomical observations, The pressures inside an exploding supernova are so great that they overcome the Pauli exclusion principle and a black hole is formed. The details are too far from the theme of this book.

Identical particles:
All electrons are identical. The Schrödinger equation with two particles has solutions which are the products of the eigenfunctions of the equation of the form:

$$\psi = u_1(p_1)u_2(p_2) \tag{14.3}$$

Because the particles are identical, if (14.3) is a solution of the Schrödinger equation, then so is:

$$\psi = u_1(p_2)u_2(p_1) \tag{14.4}$$

In which the particles have swapped places.

Symmetric and anti-symmetric wave functions:
Since the Schrödinger equation is a linear differential equation, any superposition (sum) of any solutions is also a solution. Consider:

[82] F.J. Dyson & A. Lenard: Stability of Matter J. Math. Phys. 8 423-434 (1967) &J. Math. Phys. 9 698-711 (1968)
[83] Lieb, E. H; Loss, M; Solovej, J. P. (1995) Stability of Matter in Magnetic Fields Phys. Rev. Letters 75 (6) 985-989

$$\psi_{Sym} = \frac{1}{\sqrt{2}} \left(u_1(p_1) u_2(p_2) + u_1(p_2) u_2(p_1) \right) \qquad (14.5)$$

$$\psi_{Anti-Sym} = \frac{1}{\sqrt{2}} \left(u_1(p_1) u_2(p_2) - u_1(p_2) u_2(p_1) \right) \qquad (14.6)$$

The $\frac{1}{\sqrt{2}}$ is just a normalising factor of no importance. Because the particles are identical, we can swap the particles and still have solutions of the Schrödinger equation. When we swap particles, the symmetric wave function, (14.5), is unchanged, but the anti-symmetric wave function, (14.6), is sign reversed. This is what we mean by a symmetric wave function – the non-reversal of sign under swapping of particles. This is what we mean by an anti-symmetric wave function – the reversal of sign under swapping of particles.

Now, the probability associated with these two wave functions is:

$$P = \psi^* \psi \qquad (14.7)$$

Let the two particles become similar to each other in that their four quantum numbers and their spatial position become the same. The anti-symmetric wave function then becomes:

$$\psi_{Anti-Sym} = \frac{1}{\sqrt{2}} \left(u_1(p_1) u_2(p_1) - u_1(p_1) u_2(p_1) \right) = 0 \qquad (14.8)$$

The probability of this happening is:

$$P_{Anti-Sym} = \psi^* \psi = 0.0 = 0 \qquad (14.9)$$

There is no probability of two particles described by anti-symmetric wave functions having the same quantum numbers and the same spatial position.

The symmetric wave function is different; it becomes:

$$\psi_{Sym} = \frac{1}{\sqrt{2}} \left(u_1(p_1) u_2(p_1) + u_1(p_1) u_2(p_1) \right) = 0 \qquad (14.10)$$

The probability of this happening is:

$$P_{Sym} = \psi^* \psi \neq 0 \qquad (14.11)$$

There is a probability that two particles described by symmetric wave functions will have the same quantum numbers and the same spatial position.

Within quantum theory, fermions are descibed by anti-symmetric wave functions and bosons are described by symmetric wave functions.

Summary:
Electrons, and all fermions, are described by anti-symmetric wave functions – anti-symmetric solutions of the Schrödinger equation. It could be the Pauli-Schrödinger equation or the Dirac equation.

Photons, and all bosons, are described by symmetric wave functions – symmetric solutions of the Schrödinger equation. It could be the Pauli-Schrödinger equation but not the Dirac equation. The symmetric wave functions in QFT are from the Klein-Gorden equation.

The Pauli-exclusion principle is the zero probability of two anti-symmetric wave functions simultaneously having the same quantum numbers and the same spatial position.

Chapter 15

More Upon the Dirac Equation

The Dirac spinor is taken to represent a spin one-half particle. Although, we will take this spin one-half particle to be an electron most of the time, the Dirac spinor is also taken to represent the other spin one-half particles that are the muon, the tau particle, each of the three neutrinos and each of the six quarks.

Your author has the view that the Dirac spinor is really two quaternions stacked together:

$$\Psi_{Dirac} = \begin{bmatrix} \mathbb{H}_1 \\ \mathbb{H}_2 \end{bmatrix} \tag{15.1}$$

This is eight independent variables. This is not the conventional view.

The gamma matrices as Clifford algebras:
We might expect that, since it has eight independent variables, the Dirac spinor would be presented as a 8×8 matrix, and that it might be acted upon by gamma matrices that are 8×8 matrices. The conventionally given gamma matrices are often presented as the four unit basis vectors of a 16-dimensional Clifford algebra[84] which would be sensibly written as a 16×16 matrix. If we ignore the difference between basis vectors and basis bi-vectors, we can take the gamma matrices to be elements of an 8-dimensional Clifford algebra expressed as a 8×8 matrix. The sixteen elements of the 16-dimensional Clifford algebra can be written as a set of 4×4 matrices; as such, they form a complete basis for the linear space of all 4×4 matrices.

---

[84] See: Dennis Morris The Naked Spinor

Well, now you know about the Clifford algebra view of the gamma matrices; in your author's opinion, this Clifford algebra connection is a confusing distraction.

Derivation of the Dirac equation:

We give the conventional derivation of the Dirac equation as Dirac is believed to have done it.

Historically, it was recognised that the Schrödinger equation would not alone be sufficient to deal with electrons moving at relativistic velocities. The way forward is to derive an equation similar to the Schrödinger equation from the relativistic energy momentum relation, (13.26), rather than from the classical energy momentum relation, (13.5). The Klein-Gorden equation is such a relativistic generalisation of the Schrödinger equation, and, for some time, it was thought that the Klein-Gorden equation was the only relativistic generalisation of the Schrödinger equation. Unfortunately, the Klein-Gorden equation has a single complex number as a variable. This is insufficient to represent a spin one-half particle. To deal with particles that have spin one-half, we need a field that is four independent variables; this was presumed to be a pair of complex numbers – a spinor field.

It was Dirac who discovered a second relativistic generalisation of the Schrödinger equation. Dirac required the equation to be linear like the Schrödinger equation, and, unlike the Schrödinger equation, to treat space and time on the same basis. This indicates an equation of the general form[85]:

$$H\psi = \left(\vec{\alpha}\cdot\vec{p} + \beta m\right)\psi \qquad (15.2)$$

From this equation, we get the traditional derivation of the nature of the gamma matrices – this is a mathematical classic.

---

[85] Thank you Halzen & Martin Quarks & Leptons Pg 100.

Within quantum physics, we often use the letter H to mean energy. The H stands for Hamiltonian. The four coefficients $\{\alpha_1, \alpha_2, \alpha_3, \beta\}$ must be such that the relativistic energy relation:

$$H^2 = \left(\overrightarrow{p}^2 + m^2\right) \tag{15.3}$$

is satisfied. We cannot be sure that the coefficients will multiplicatively commute with each other; indeed, we would not have to play around with the coefficients for long before we realised that they must not commute with each other. In the next calculation, we bear this non-commutativity in mind. From (15.2), we have:

$$H^2 = \left(\alpha_i P_i + \beta m\right)\left(\alpha_i P_i + \beta m\right)$$
$$= \alpha_i^2 P_i^2 + \left(\alpha_i \alpha_j + \alpha_j \alpha_i\right) P_i P_j + \left(\alpha_i \beta + \beta \alpha_i\right) P_i m + \beta^2 m^2$$
$$\tag{15.4}$$

To satisfy the relativistic energy relation, (15.3), we need:

$$\alpha_1^2 = 1, \quad \alpha_2^2 = 1, \quad \alpha_3^2 = 1, \quad \beta^2 = 1$$
$$\alpha_i \alpha_j = -\alpha_j \alpha_i, \qquad\qquad \alpha_i \beta = -\beta \alpha_i \tag{15.5}$$

This is very similar to the basis vectors of a Clifford algebra, but this is not a 4-dimensional Clifford algebra. There are no 4-dimensional Clifford algebras with four square roots of plus one. Three 8-dimensional Clifford algebras have four square roots of plus unity.

We see that the coefficients of (15.2) do not commute, (15.5). Non-commutation is expressed as matrices; the coefficients cannot be real or complex numbers; the smallest matrices which can hold the commutation relations of the coefficients, (15.5), are 4×4 matrices. This is what leads to the Dirac spinor being a 4-component object – to fit with multiplication by 4×4 matrices. Of course, the Dirac spinor is really an 8-component object disguised as a 4-component object.

There is a little problem with the equation presented above, (15.2). The amount of mass, β, is a non-commutative matrix; it ought to be a real number. We can solve this problem by multiplying from the left by β:

$$\beta H \psi = (\beta \vec{\alpha} \cdot \vec{p} + m) \psi \tag{15.6}$$

Putting the quantum mechanical operators in place of the energy and momentum:

$$i\beta \frac{\partial}{\partial t} \psi = \left(-i\beta \vec{\alpha} \cdot \frac{\overrightarrow{\partial}}{\partial x} + m \right) \psi \tag{15.7}$$

We re-label the matrices $\beta = \gamma^0$, $\beta \alpha_1 = \gamma^1$, $\beta \alpha_2 = \gamma^2$, $\beta \alpha_3 = \gamma^3$, and we have the Dirac equation:

$$i\gamma^\mu \partial_\mu \psi - m\psi = 0 \tag{15.8}$$

Since $\gamma^0 = \beta$, we have $\gamma_0^2 = +1$. Now:

$$(\beta \alpha_i)^2 = \beta \alpha_i \beta \alpha_i = -\beta \alpha_i \alpha_i \beta = -\beta\beta = -1 \tag{15.9}$$

and so we have:

$$\gamma_0^2 = +1, \qquad \gamma_1^2 = -1, \qquad \gamma_2^2 = -1, \qquad \gamma_3^2 = -1 \tag{15.10}$$

We have shown that, to satisfy the relativistic energy momentum relation, the gamma matrices must be of the form (15.10).

The adjoint Dirac spinor:
We have the Dirac equation above, (15.8). We expand the notation a little:

$$i\gamma^0 \frac{\partial}{\partial t} \psi + i\gamma^\mu \frac{\partial}{\partial x^\mu} \psi - m\psi = 0 \tag{15.11}$$

We will need the hermitian conjugate of the Dirac equation, this is:

$$-i\frac{\partial}{\partial t}\psi^\dagger \gamma^0 - i\frac{\partial}{\partial x^\mu}\psi^\dagger (-\gamma^\mu) - m\psi^\dagger = 0 \tag{15.12}$$

but this is not of the same form as the Dirac equation; we have a minus sign in the wrong place. However, if instead of using $\psi^\dagger$, we had used $\overline{\psi} = \psi^\dagger \gamma^0$, then we would have the correct form for the hermitian conjugate Dirac equation, which is:

$$i\frac{\partial}{\partial x^\mu}\overline{\psi}\gamma^\mu + m\overline{\psi} = 0 \tag{15.13}$$

We therefore conclude that the correct adjoint of the Dirac spinor is:

$$\overline{\psi} = \psi^\dagger \gamma^0 \tag{15.14}$$

The reader might think this is a little contrived. This contrivance works, but, as we will see later, this contrivance is necessary because the Dirac spinor is written as two pairs of complex numbers rather than as two quaternions. If we had taken the Dirac spinor to be two quaternions in normal notation, then there would be no need of his contrivance.

The continuity equation:

Multiplying the Dirac equation, (15.8), from the left by $\overline{\psi}$ and multiplying the hermitian conjugate Dirac equation, (15.13), from the right by ψ and adding gives:

$$\overline{\psi}\gamma^\mu\partial_\mu\psi + \left(\partial_\mu\overline{\psi}\right)\gamma^\mu\psi = \partial_\mu\left(\overline{\psi}\gamma^\mu\psi\right) = 0 \tag{15.15}$$

The continuity equation is $\partial_\mu j^\mu = 0$. We have a continuity equation if we put:

$$j^\mu = \overline{\psi}\gamma^\mu\psi \tag{15.16}$$

If we interpret this to be the probability and flux densities with j^0 being the probability flux density, we have:

$$\rho = j^0 = \overline{\psi}\gamma^0\psi = \psi^\dagger\psi \tag{15.17}$$

This, (15.17), is:

$$
\begin{bmatrix} a-ib & c-id & e-if & g-ih \end{bmatrix}
\begin{bmatrix} a+ib \\ c+id \\ e+if \\ g+ih \end{bmatrix} \tag{15.18}
$$

$$
= \left(a^2 + b^2 + c^2 + d^2\right) + \left(e^2 + f^2 + g^2 + h^2\right)
$$

This is the probability. It is positive definite.

It is also the sum of two quaternion norms. If we had written:

$$
\begin{bmatrix} a+ib \\ c+id \end{bmatrix} \equiv
\begin{bmatrix}
a & b & c & d \\
-b & a & -d & c \\
-c & d & a & -b \\
-d & -c & b & a
\end{bmatrix}
\quad \& \quad
\begin{bmatrix} a+ib \\ c+id \end{bmatrix}^{\dagger} \equiv
\begin{bmatrix}
a & -b & -c & -d \\
b & a & d & -c \\
c & -d & a & b \\
d & c & -b & a
\end{bmatrix}
\tag{15.19}
$$

and similarly the other part of the Dirac spinor, we would have come to the same probability expression for $\psi^{\dagger}\psi$. Looking carefully at the above, we see how the contrived Dirac adjoint would have been avoided if we used quaternions.

The probability and flux density is actually interpreted to be the charge current density, and so we need to modify (15.16) by multiplying it by the charge of the electron:

$$
j^{\mu} = -e\overline{\psi}\gamma^{\mu}\psi \tag{15.20}
$$

The expression j^{μ} is a 4-vector ($\mu = 0, 1, 2, 3$) and is taken to be the electron current density.

Bi-linear covariants:

From the Dirac adjoint spinor, $\overline{\Psi}$, the Dirac spinor, Ψ, and the gamma matrices, we can form various kinds of mathematical objects. These different objects are known as the bi-linear covariants of the

Dirac spinor. We have already met one of these bi-linear covariants above, (15.18). Between them, the bi-linear covariants describe the state of the electron. Two bi-linear covariants are scalars (just numbers). There are two sets of four bi-linear covariants that are two vectors – one is an axial vector. There is one set of six bi-linear covariants that form an anti-symmetric tensor. Altogether, this is sixteen bi-linear covariants that describe the state of the electron.

The sixteen bi-linear covariants correspond to the basis elements of a sixteen dimensional Clifford algebra. There are five 16-dimensional Clifford algebras; however, seen as division algebras, these five 16-dimensional Clifford algebras form only two distinct 16-dimensional division algebras. The reader might wonder why one of the five, or one of the two, 16-dimensional algebras is preferred by nature. Your author thinks this Clifford algebra stuff is meaningless distraction.

The actual bi-linear covariants:
With:

$$\Psi = \begin{bmatrix} a+ib \\ c+id \\ e+if \\ g+ih \end{bmatrix} \qquad (15.21)$$

and the gamma matrices (13.40) & (13.41), we have, a scalar:

$$\overline{\Psi}\Psi = a^2 + b^2 + c^2 + d^2 - e^2 - f^2 - g^2 - h^2 \qquad (15.22)$$

A vector; one component for each μ :

$$\Psi\gamma^{\mu}\Psi = \begin{bmatrix} a^2 + b^2 + c^2 + d^2 + e^2 + f^2 + g^2 + h^2 \\ 2(ag + bh + ce + df) \\ 2(ah - bg - cf + ed) \\ 2(ae + bf - cg - dh) \end{bmatrix} \qquad (15.23)$$

The anti-symmetric tensor is, $\sigma^{\mu\nu} = \dfrac{i}{2}\left(\gamma^{\mu}\gamma^{\nu} - \gamma^{\nu}\gamma^{\mu}\right)$:

$$\overline{\Psi}\sigma^{\mu\nu}\Psi =$$

$$
\begin{bmatrix}
0 & 2\begin{pmatrix} -ah+bg \\ -cf+de \end{pmatrix} & 2\begin{pmatrix} ag+bh \\ -ce-df \end{pmatrix} & 2\begin{pmatrix} -af+be \\ +ch-dg \end{pmatrix} \\
2\begin{pmatrix} ah-bg \\ +cf-de \end{pmatrix} & 0 & \begin{matrix} a^2+b^2-c^2-d^2 \\ -e^2-f^2+g^2+h^2 \end{matrix} & 2\begin{pmatrix} -ad+bc \\ +eh-fg \end{pmatrix} \\
2\begin{pmatrix} -ag-bh \\ +ce+df \end{pmatrix} & \begin{matrix} -a^2-b^2+c^2+d^2 \\ +e^2+f^2-g^2-h^2 \end{matrix} & 0 & 2\begin{pmatrix} ac+bd \\ -eg-fh \end{pmatrix} \\
2\begin{pmatrix} af-be \\ -ch+dg \end{pmatrix} & 2\begin{pmatrix} ad-bc \\ -eh+fg \end{pmatrix} & 2\begin{pmatrix} -ac-bd \\ +eg+fh \end{pmatrix} & 0
\end{bmatrix}
$$

$$(15.24)$$

An axial vector, $\gamma^5 = \gamma^0\gamma^1\gamma^2\gamma^3$:

$$
\overline{\Psi}\gamma^5\gamma^{\mu}\Psi = \begin{bmatrix}
-2(ae+bf+cg+dh) \\
-2(ac+bd+eg+fh) \\
-2(ad-bc+eh-fg) \\
-a^2-b^2+c^2+d^2-e^2-f^2+g^2+h^2
\end{bmatrix}
\qquad (15.25)
$$

And a pseudo-scalar:

$$\overline{\Psi}\gamma^5\Psi = 2(af-be+ch-dg) \qquad (15.26)$$

The integrals over space-time of the bi-linear covariants give the expectation values of the physical observable properties of the electron.

The two scalars were combined together by de Broglie to form[86]:

$$\Omega = \overline{\Psi}\Psi + \overline{\Psi}\gamma^5\Psi\gamma^5 \qquad (15.27)$$

---

[86] Pertti Lounesto Clifford Algebras and Spinors pg138

The first component of the vector, integrated over space-time, gives the probability of finding the electron in the domain of integration. The other vector components are the current probability.

The anti-symmetric tensor is interpreted as the electromagnetic moment density.

The axial vector gives the direction of the spin of the electron.

Weyl spinors:

A proper mathematician would insist that equations have to be written within a division algebra. This means that every object, gamma matrices, spinor,… must be an element of that division algebra. Put simply, every object in a quaternion equation should be a quaternion. A pair of quaternions is not a division algebra; therefore, to be pedantic, the Dirac equation is a mathematical nonsense. "The Dirac equation works", screams the physicist. We have a problem.

The Dirac spinor is a set of four complex numbers, but it is often presented, as we have presented it above, as two pairs of complex numbers:

$$\Psi_{Dirac} = \begin{bmatrix} a+ib \\ c+id \\ e+if \\ g+ih \end{bmatrix} \rightarrow \begin{bmatrix} \begin{bmatrix} a+ib \\ c+id \end{bmatrix} \\ \begin{bmatrix} e+if \\ g+ih \end{bmatrix} \end{bmatrix} = \begin{bmatrix} \eta \\ \chi \end{bmatrix} \qquad (15.28)$$

The 2-component entities, $\{\eta, \chi\}$ are called Weyl spinors after Hermann Weyl (1885-1955).

Such decomposition makes sense because η and χ transform independently under Lorentz transformations; this means the components of the two Weyl spinors do not get mixed together when we rotate the co-ordinate system in 2-dimensional space-time (a

Lorentz boost) or when we rotate the co-ordinate system in 2-dimensional Euclidean space.

Technically, the Dirac spinor is a reducible representation of the Lorentz group whereas the Weyl spinors are irreducible representations of the Lorentz group. Pictorially, a Dirac spinor is a 8×8 matrix with the two Weyl spinors on the leading diagonal, and this 8×8 matrix is acted upon by a 4×4 Lorentz group rotation matrix[87]:

$$[R]\begin{bmatrix} [W] & 0 \\ 0 & [W] \end{bmatrix} \qquad (15.29)$$

The 4×4 rotation matrix, $[R]$, has to act separately on each of the 4×4 $[W]$ matrices[88]. Effectively, the 8×8 Dirac spinor matrix can be reduced to two separate 4×4 matrices.

There is ambiguity around how physicists define Weyl spinors. We quote Patrick Labelle[89]. *"... some references restrict the term Weyl spinors to massless spinors, in which case Weyl spinors are always eigenstates of the helicity operator. Others define Weyl spinors as being the chirality eigenstates, which are irreducible representations of the Lorentz group. With the first definition, it is helicity that is used as the defining property of Weyl spinors, whereas in the second convention, it is behaviour under Lorentz transformations that is judged as being more fundamental. For the proponents of the first definition, massive Weyl fermions is an oxymoron, whereas the other camp has no problem with that."*

In passing, we should add that there are Majorana spinors. These are comprised of four complex numbers (eight independent variables).

---

[87] There are six 4×4 rotation matrices in the Lorentz group. They correspond to the six 2-dimensional rotations in our 4-dimensional space-time – three Euclidean rotations and three 2-dimensional space-time rotations.

[88] The London Mathematical Society would crucify me for putting it this way, but it does show what the technical stuff means.

[89] Patrick Labelle: Supersymmetry DeMystified Page 73

The Dirac equation again:

There is another way to approach the derivation of the Dirac equation[90].

We take it that particles moving at the speed of light are massless. In this case, the relativistic energy momentum relation, (13.26), becomes:

$$E^2 = p^2 \qquad (15.30)$$

We associate the angular frequency, ω, of a wave with the energy of the wave and the wave number, k, with the momentum of the wave. Thus, at the speed of light, for massless particles, we have:

$$\omega^2 = k^2$$
$$\omega = \pm k \qquad (15.31)$$

We then assume an electron wave of the form:

$$\psi = e^{i(kx-\omega t)} \equiv \begin{bmatrix} \cos(kx-\omega t) & \sin(kx-\omega t) \\ -\sin(kx-\omega t) & \cos(kx-\omega t) \end{bmatrix} \qquad (15.32)$$

We have:

$$\frac{\partial \psi}{\partial t} = -i\omega e^{i(kx-\omega t)} \qquad \frac{\partial \psi}{\partial x} = ike^{i(kx-\omega t)} \qquad (15.33)$$

Since, at the velocity of light, $\omega = \pm k$, from (15.33), we get:

$$\frac{\partial \psi}{\partial t} = -\frac{\partial \psi}{\partial x}$$
$$\frac{\partial \psi}{\partial t} = \frac{\partial \psi}{\partial x} \qquad (15.34)$$

---

[90] For this approach, we thank Leonard Susskind of Stanford University for his lectures on 'Basic Concepts'. This is lecture 5 at 1 hr 45 mins into the lecture and the following lecture 6. These lectures are freely available on the internet through iTunes.

One of this equations, (15.34), represents a rightward-moving electron wave, and the other equation represents a leftward moving electron wave – it does not matter which is which. We represent the leftward moving electron wave by ψ_L and the rightward moving electron wave by ψ_R. This leads to:

$$\frac{\partial \psi_R}{\partial x} - \frac{\partial \psi_R}{\partial t} = 0$$
$$\frac{\partial \psi_L}{\partial x} + \frac{\partial \psi_L}{\partial t} = 0 \tag{15.35}$$

These are two massless electron fields moving left and right respectively at the speed of light. Yes, I know there are no massless electrons, but stick with the argument for now.

We now introduce some notation; this is not physics, this is no more than notation. We fit the two electron fields together into a vector:

$$\Psi = \begin{bmatrix} \psi_R \\ \psi_L \end{bmatrix} \tag{15.36}$$

The two equations, (15.35), now become:

$$\frac{\partial}{\partial t}\begin{bmatrix} \psi_R \\ \psi_L \end{bmatrix} = -\alpha \frac{\partial}{\partial x}\begin{bmatrix} \psi_R \\ \psi_L \end{bmatrix} \qquad : \quad \alpha = \begin{bmatrix} 1 & 0 \\ 0 & -1 \end{bmatrix}$$
$$\frac{\partial}{\partial t}\begin{bmatrix} \psi_R \\ \psi_L \end{bmatrix} = -\frac{\partial}{\partial x}\begin{bmatrix} \psi_R \\ -\psi_L \end{bmatrix} \tag{15.37}$$

Differentiation with respect to both $\{t,x\}$ will fetch a i into the equation, or, if the reader prefers, we simply multiply by i to give:

$$i\frac{\partial}{\partial t}\begin{bmatrix} \psi_R \\ \psi_L \end{bmatrix} = -i\frac{\partial}{\partial x}\begin{bmatrix} \psi_R \\ -\psi_L \end{bmatrix} \tag{15.38}$$

We now seek a massive electron wave. For a massive electron wave, we have:

$$E^2 = m^2 + p^2$$
$$\omega^2 = m^2 + k^2 \tag{15.39}$$

We add the mass into the equation (15.37) as:

$$\omega = \alpha k + \beta m \tag{15.40}$$

Then, as we did above, from (15.2) to (15.5), we take the square of (15.40) and we come to the relations:

$$\alpha^2 = 1, \quad \beta^2 = 1, \quad \alpha\beta = -\beta\alpha \tag{15.41}$$

Note that these are a different set of matrices from the ones given above from (15.2) to (15.5); here, we have:

$$\alpha = \begin{bmatrix} 1 & 0 \\ 0 & -1 \end{bmatrix} \qquad \beta = \begin{bmatrix} 0 & 1 \\ 1 & 0 \end{bmatrix} \tag{15.42}$$

This difference is no more than a different choice of basis for the gamma matrices. Equation, (15.38), now becomes massive as:

$$i\frac{\partial}{\partial t}\begin{bmatrix} \psi_R \\ \psi_L \end{bmatrix} = -i\frac{\partial}{\partial x}\begin{bmatrix} \psi_R \\ -\psi_L \end{bmatrix} + \beta\begin{bmatrix} \psi_R \\ \psi_L \end{bmatrix}m$$
$$= -i\frac{\partial}{\partial x}\begin{bmatrix} \psi_R \\ -\psi_L \end{bmatrix} + \begin{bmatrix} \psi_L \\ \psi_R \end{bmatrix}m \tag{15.43}$$

We see that the matrix β has flipped the $\psi_L \& \psi_R$ positions. Written as two equations, this is the two coupled differential equations:

$$i\left(\frac{\partial}{\partial t} + \frac{\partial}{\partial x}\right)\psi_R = m\psi_L$$
$$i\left(\frac{\partial}{\partial t} - \frac{\partial}{\partial x}\right)\psi_L = m\psi_R \tag{15.44}$$

It seems that the mass term couples the leftward moving electron field, ψ_L, and the rightward moving electron field, ψ_R, together. We thus have a view of the massive electron being a superposition of two massless electron fields. The superposition is formed in two ways:

$$\psi_+ = \psi_L + \psi_R$$
$$\psi_- = \psi_L - \psi_R \qquad (15.45)$$

The first of these super positions corresponds to:

$$i\frac{\partial}{\partial t}\psi_+ = m\psi_+ \qquad : \qquad \omega = +m \qquad (15.46)$$

This has positive energy and represents an electron. The second superposition corresponds to:

$$i\frac{\partial}{\partial t}\psi_- = -m\psi_- \qquad : \qquad \omega = -m \qquad (15.47)$$

This has negative energy and represents a positron.

Massless fermions:
Let us take the Dirac equation and set the mass to zero:

$$\gamma^\mu\partial_\mu\psi = 0 \qquad (15.48)$$

This equation is taken to describe massless fermions. The only massless fermions are neutrinos, and we are not sure that neutrinos are massless. Thus, ignoring the possibility that neutrinos might have mass, the massless Dirac equation, (15.48), is taken to describe neutrinos. In a later chapter, we will derive a massless neutrino-field alongside a massive electron-field, and so we have no need of the assumption that (15.48) describes neutrinos.

More of the Dirac equation:
We begin with the Dirac equation, (15.8):

$$i\gamma^\mu\partial_\mu\psi - m\psi = 0 \qquad (15.49)$$

And we use tha gamma matrices given above, (13.40) & (13.41): This gives the Dirac equation as:

$$i\frac{\partial}{\partial t}\begin{bmatrix}\psi_1\\\psi_2\\-\psi_3\\-\psi_4\end{bmatrix}+i\frac{\partial}{\partial x}\begin{bmatrix}\psi_4\\\psi_3\\-\psi_2\\-\psi_1\end{bmatrix}-\frac{\partial}{\partial y}\begin{bmatrix}-\psi_4\\\psi_3\\\psi_2\\-\psi_1\end{bmatrix}+i\frac{\partial}{\partial z}\begin{bmatrix}\psi_3\\-\psi_4\\-\psi_1\\-\psi_2\end{bmatrix}-m\begin{bmatrix}\psi_1\\\psi_2\\\psi_3\\\psi_4\end{bmatrix}=0$$

$$(15.50)$$

The orders of the ∂_x & ∂_y are a matter of basis, and the above could be written as:

$$i\frac{\partial}{\partial t}\begin{bmatrix}\psi_1\\\psi_2\\-\psi_3\\-\psi_4\end{bmatrix}+i\frac{\partial}{\partial x}\begin{bmatrix}\psi_3\\\psi_4\\-\psi_1\\-\psi_2\end{bmatrix}-\frac{\partial}{\partial y}\begin{bmatrix}\psi_3\\-\psi_4\\-\psi_1\\\psi_2\end{bmatrix}+i\frac{\partial}{\partial z}\begin{bmatrix}\psi_3\\-\psi_4\\-\psi_1\\-\psi_2\end{bmatrix}-m\begin{bmatrix}\psi_1\\\psi_2\\\psi_3\\\psi_4\end{bmatrix}=0$$

$$(15.51)$$

We see the four different complex numbers, ψ_i, come in two pairs, $\{\psi_1,\psi_2\}$ & $\{\psi_3,\psi_4\}$. These pairs correspond to the ψ_+ & ψ_- given above.

Solutions of the Dirac equation:
Solving the Dirac equation is based on taking the Dirac spinor of four complex numbers to be two pairs of complex numbers:

$$\begin{bmatrix}a+ib\\c+id\\e+if\\g+ih\end{bmatrix}\rightarrow\begin{bmatrix}\begin{bmatrix}a+ib\\c+id\end{bmatrix}\\\begin{bmatrix}e+if\\g+ih\end{bmatrix}\end{bmatrix}$$

$$(15.52)$$

This leads after some manipulation[91] to the energies of the particles described by the Dirac equation being:

---

[91] See David McMahon Quantum Field Theory DeMystified pg 99

$$E = \omega_k = k_0 = \pm\sqrt{k^2 + m^2} \tag{15.53}$$

The particles described by the Dirac equation are spin one half particles, fermions, because a pair of complex numbers represents such particles. Thus we have electrons with both positive and negative energies. The positive energy electrons are taken to be electrons, and the negative energy electrons are taken to be positrons.

Free particle solutions of the Dirac equation:
For a particle at rest, we can disregard the spatial derivatives – they are all zero. The Dirac equation then becomes:

$$i\gamma^0 \frac{\partial \psi}{\partial t} - m\psi = 0$$

$$\psi = \begin{bmatrix} u \\ v \end{bmatrix}, \quad \gamma^0 = \begin{bmatrix} 1 & 0 \\ 0 & -1 \end{bmatrix} \tag{15.54}$$

This is two differential equations:

$$i\frac{\partial u}{\partial t} = mu$$

$$i\frac{\partial v}{\partial t} = -mv \tag{15.55}$$

With solutions:

$$u(t) = u(0)e^{-imt}$$

$$v(t) = v(0)e^{imt} \tag{15.56}$$

These two solutions are just clockwise rotation in the complex plane and anti-clockwise rotation in the complex plane:

$$u(t) = \begin{bmatrix} u(0) & 0 \\ 0 & u(0) \end{bmatrix} \begin{bmatrix} \cos(mt) & -\sin(mt) \\ \sin(mt) & \cos(mt) \end{bmatrix}$$

$$v(t) = \begin{bmatrix} v(0) & 0 \\ 0 & v(0) \end{bmatrix} \begin{bmatrix} \cos(mt) & \sin(mt) \\ -\sin(mt) & \cos(mt) \end{bmatrix} \tag{15.57}$$

The two solutions are taken to be particle and anti-particle. Thus the difference between an electron and a positron is that one is the conjugate of the other. Remember that, using Noether's theorem, we associated rotation in the complex plane with electric charge.

The Dirac Equation Done Differently

We begin with the relativistic energy momentum relation:

$$E^2 = p^2 + m^2 \qquad (16.1)$$

Although momentum is a vector, the momentum squared:

$$p^2 = \vec{p} \cdot \vec{p} = p_x p_x + p_y p_y + p_z p_z \qquad (16.2)$$

is a scalar, and so the above equation, (16.1), is a scalar equation – just real numbers. This equation comes straight out of special relativity.

We rewrite the relativistic energy momentum equation as 4×4 matrices:

$$\begin{bmatrix} E^2 & 0 & 0 & 0 \\ 0 & E^2 & 0 & 0 \\ 0 & 0 & E^2 & 0 \\ 0 & 0 & 0 & E^2 \end{bmatrix} = \begin{bmatrix} m^2 + p^2 & 0 & 0 & 0 \\ 0 & m^2 + p^2 & 0 & 0 \\ 0 & 0 & m^2 + p^2 & 0 \\ 0 & 0 & 0 & m^2 + p^2 \end{bmatrix}$$
$$(16.3)$$

We now take the square root of both sides of the matrix equation, (16.3). The square root of the LHS of (16.3) is simply:

$$\pm \begin{bmatrix} E & 0 & 0 & 0 \\ 0 & E & 0 & 0 \\ 0 & 0 & E & 0 \\ 0 & 0 & 0 & E \end{bmatrix} \qquad (16.4)$$

However, the RHS is the square of a quaternion. This is a quaternion multiplied by its conjugate:

$$\begin{bmatrix} m & p_x & p_y & p_z \\ -p_x & m & -p_z & p_y \\ -p_y & p_z & m & -p_x \\ -p_z & -p_y & p_x & m \end{bmatrix} \begin{bmatrix} m & -p_x & -p_y & -p_z \\ p_x & m & p_z & -p_y \\ p_y & -p_z & m & p_x \\ p_z & p_y & -p_x & m \end{bmatrix}$$

$$= \begin{bmatrix} m^2 + p^2 & 0 & 0 & 0 \\ 0 & m^2 + p^2 & 0 & 0 \\ 0 & 0 & m^2 + p^2 & 0 \\ 0 & 0 & 0 & m^2 + p^2 \end{bmatrix} \quad (16.5)$$

$$m^2 + p_x^2 + p_y^2 + p_z^2 = m^2 + p^2$$

The square root equation now becomes two equations:

$$\begin{bmatrix} E & 0 & 0 & 0 \\ 0 & E & 0 & 0 \\ 0 & 0 & E & 0 \\ 0 & 0 & 0 & E \end{bmatrix} = \begin{bmatrix} m & p_x & p_y & p_z \\ -p_x & m & -p_z & p_y \\ -p_y & p_z & m & -p_x \\ -p_z & -p_y & p_x & m \end{bmatrix} \quad (16.6)$$

and:

$$\begin{bmatrix} E & 0 & 0 & 0 \\ 0 & E & 0 & 0 \\ 0 & 0 & E & 0 \\ 0 & 0 & 0 & E \end{bmatrix} = \begin{bmatrix} m & -p_x & -p_y & -p_z \\ p_x & m & p_z & -p_y \\ p_y & -p_z & m & p_x \\ p_z & p_y & -p_x & m \end{bmatrix} \quad (16.7)$$

We extract the matrices from the quaternion matrix equation (16.6):

$$\lambda^0 = \begin{bmatrix} 1 & 0 & 0 & 0 \\ 0 & 1 & 0 & 0 \\ 0 & 0 & 1 & 0 \\ 0 & 0 & 0 & 1 \end{bmatrix} \qquad \lambda^1 = \begin{bmatrix} 0 & 1 & 0 & 0 \\ -1 & 0 & 0 & 0 \\ 0 & 0 & 0 & -1 \\ 0 & 0 & 1 & 0 \end{bmatrix} \quad (16.8)$$

$$\lambda^2 = \begin{bmatrix} 0 & 0 & 1 & 0 \\ 0 & 0 & 0 & 1 \\ -1 & 0 & 0 & 0 \\ 0 & -1 & 0 & 0 \end{bmatrix} \qquad \lambda^3 = \begin{bmatrix} 0 & 0 & 0 & 1 \\ 0 & 0 & -1 & 0 \\ 0 & 1 & 0 & 0 \\ -1 & 0 & 0 & 0 \end{bmatrix} \qquad (16.9)$$

This gives:

$$\lambda^0 E = \lambda^1 p_x + \lambda^2 p_y + \lambda^3 p_z + \lambda^0 m \qquad (16.10)$$

If we follow the standard derivation of the Dirac equation, we might now substitute into the square root equations, (16.6) or (16.7) the momentum and energy operators:

$$E = i\hbar \frac{\partial}{\partial t} \qquad \& \qquad p = -i\hbar \frac{\partial}{\partial x} \qquad (16.11)$$

We set $\hbar = 1$, and get:

$$\lambda^0 i \frac{\partial}{\partial t} = -\lambda^1 i \frac{\partial}{\partial x} - \lambda^2 i \frac{\partial}{\partial y} - \lambda^3 i \frac{\partial}{\partial z} + \lambda^0 m$$

$$i\lambda^\mu \frac{\partial}{\partial x^\mu} - \lambda^0 m = 0 \qquad (16.12)$$

The λ^0 matrix is just the real number one; technically, it is algebraically isomorphic to the number one, and so we can forget it as a coeficient of the mass term. We give the operators in equation (16.12) a field to operate upon, $\psi(t,x)$, and we have:

$$i\lambda^\mu \frac{\partial}{\partial x^\mu} \psi(t,x) - m\psi(t,x) = 0 \qquad (16.13)$$

The matrices λ^i satisfy all the conditions required of the gamma matrices, (13.38) & (13.39). We have the Dirac equation – not quite. This, (16.13), is a quaternion equation. The field, $\psi(t,x)$, must be a quaternion, not a Dirac spinor.

We have another equation using the other square root, the conjugate quaternion, and we are led to the equation:

$$-i\lambda^0 \frac{\partial}{\partial t}\psi(t,x)+i\lambda^\mu \frac{\partial}{\partial x^\mu}\psi(t,x)+m\psi(t,x)=0 \quad (16.14)$$

This is not quite the conjugate of the equation (16.13); there is a minus sign before the λ^0 term.

However, we have made an error. Differentiation within the quaternions is non-commutative. We really ought to use the $SU(2)$ differential operators rather than the operators in (16.11). We will address this shortly.

The anti-quaternion square root:
There are two quaternion algebras, the quaternions and the anti-quaternions[92]. The anti-quaternions can provide a root to the relativistic energy momentum relation. We have:

$$\begin{bmatrix} m & p_x & p_y & p_z \\ -p_x & m & p_z & -p_y \\ -p_y & -p_z & m & p_x \\ -p_z & p_y & -p_x & m \end{bmatrix}\begin{bmatrix} m & -p_x & -p_y & -p_z \\ p_x & m & -p_z & p_y \\ p_y & p_z & m & -p_x \\ p_z & -p_y & p_x & m \end{bmatrix}$$

$$= \begin{bmatrix} m^2+p^2 & 0 & 0 & 0 \\ 0 & m^2+p^2 & 0 & 0 \\ 0 & 0 & m^2+p^2 & 0 \\ 0 & 0 & 0 & m^2+p^2 \end{bmatrix} \quad (16.15)$$

$$m^2+p_x^2+p_y^2+p_z^2=m^2+p^2$$

The commutation relations of the anti-quaternions are in the opposite direction to the quaternions, see (12.12). Thus we have a right-chiral electron and right-chiral positron as well as a left-chiral electron and a left-chiral positron.

---

[92] The anti-quaternions are poorly named. It would be better to refer to the two quaternion algebras as the left-chiral quaternions and the right-chiral quaternions. The poor nomenclature is historical.

Majorana solutions:

We mentioned earlier, (13.46), that, if we do not require the gamma matrices to form a division algebra, then we can choose the gamma matrices to be entirely imaginary. This would lead to the conventional Dirac equation being a real equation with real solutions. Since an anti-particle is just the conjugate of the spinor (a pair of complex numbers in the conventional view, a quaternion in your author's view) real solutions of the Dirac equation are particles, called Majorana particles, which are their own anti-particles.

If we insist upon the gamma matrices being the unit elements of a division algebra, we have only two possible sets of gamma matrices – the two quaternion algebras. Niether of these two sets of gamma matrices are entirely imaginary, and so we do not have real solutions to the quaternion Dirac equation, and so we do not have Majorana particles.

The differential operators:

The differential operators given above, (16.11), are based on differentiation within the Euclidean complex numbers[93]. The Euclidean complex numbers are a commutative division algebra.

Let us look at the Euclidean complex numbers, $\mathbb{C}_\lambda$. We have:

$$\begin{bmatrix} 0 & 1 \\ -\lambda & 0 \end{bmatrix}^2 = \lambda \begin{bmatrix} -1 & 0 \\ 0 & -1 \end{bmatrix} \tag{16.16}$$

We will differentiate a complex function with respect to the imaginary axis:

---

[93] See: Dennis Morris The Physics of Empty Space pg 80

$$\frac{\partial \begin{bmatrix} f(a,b) & g(a,b) \\ -\lambda g(a,b) & f(a,b) \end{bmatrix}}{\partial \begin{bmatrix} 0 & b \\ -\lambda b & 0 \end{bmatrix}} = \frac{1}{\begin{bmatrix} 0 & 1 \\ -\lambda & 0 \end{bmatrix}} \frac{\partial \begin{bmatrix} f(a,b) & g(a,b) \\ -\lambda g(a,b) & f(a,b) \end{bmatrix}}{\partial \begin{bmatrix} b & 0 \\ 0 & b \end{bmatrix}}$$

$$= \begin{bmatrix} 0 & -\dfrac{1}{\lambda} \\ 1 & 0 \end{bmatrix} \begin{bmatrix} \dfrac{\partial f}{\partial b} & \dfrac{\partial g}{\partial b} \\ -\lambda \dfrac{\partial g}{\partial b} & \dfrac{\partial f}{\partial b} \end{bmatrix} \qquad (16.17)$$

$$= \begin{bmatrix} 0 & -1 \\ \lambda & 0 \end{bmatrix} \begin{bmatrix} \dfrac{1}{\lambda} & 0 \\ 0 & \dfrac{1}{\lambda} \end{bmatrix} \begin{bmatrix} \dfrac{\partial f}{\partial b} & \dfrac{\partial g}{\partial b} \\ -\lambda \dfrac{\partial g}{\partial b} & \dfrac{\partial f}{\partial b} \end{bmatrix}$$

We see that the process of differentiation by an imaginary variable necessitates multiplication by the negative of the imaginary unit, $-i_\lambda$ and by the inverse of the scaling parameter, $\dfrac{1}{\lambda}$. We might, in other notation, write this differentiation operation as:

$$-i_\lambda \frac{1}{\lambda} \frac{\partial}{\partial b} \qquad (16.18)$$

Within quantum mechanics, we have the momentum operator:

$$p_x = -ih \frac{\partial}{\partial x} \qquad (16.19)$$

We see that:

$$h \equiv \frac{1}{\lambda} \qquad (16.20)$$

Since we are working with quaternions, which are non-commutative, we ought to use non-commutative $SU(2)$ differentiation operators[94]. The non-commutative quaternion differentiation operator is:

$$\begin{bmatrix} \partial t & -\partial x & -\partial y & -\partial z \\ \partial x & \partial t & \partial z & -\partial y \\ \partial y & -\partial z & \partial t & \partial x \\ \partial z & \partial y & -\partial x & \partial t \end{bmatrix} \quad (16.21)$$

This acts on a quaternion to the left, which we call d_L, and on a quaternion to the right, which we call d_R. Non-commutative differentiation leads to two quaternion fields defined by:

$$E_{\mathbb{H}} = \frac{1}{2}(d_L + d_R)$$
$$B_{\mathbb{H}} = \frac{1}{2}(d_L - d_R) \quad (16.22)$$

We do not need to include the $i = \sqrt{-1}$ because this does nothing more than choose the imaginary parts of the matrix as examination of (16.17) demonstrates.

We need to substitute these differentials into the basic equation given above, (16.10). We have an unfortunate conflict of notation, and so we will rename the energy with an H. This makes the basic equation, (16.10), into:

$$\lambda^0 H = \lambda^1 p_x + \lambda^2 p_y + \lambda^3 p_z + m \quad (16.23)$$

We put:

---

[94] See Dennis Morris *The Physics of Empty Space*. Non-commutative differentiation is presented in tedious detail in this book. It is also presented more tersely in the book Dennis Morris: *Quaternions*.

$$H = \begin{bmatrix} \partial t & 0 & 0 & 0 \\ 0 & \partial t & 0 & 0 \\ 0 & 0 & \partial t & 0 \\ 0 & 0 & 0 & \partial t \end{bmatrix} \qquad P_x = \begin{bmatrix} 0 & -\partial x & 0 & 0 \\ \partial x & 0 & 0 & 0 \\ 0 & 0 & 0 & \partial x \\ 0 & 0 & -\partial x & 0 \end{bmatrix}$$

$$P_y = \begin{bmatrix} 0 & 0 & -\partial y & 0 \\ 0 & 0 & 0 & -\partial y \\ \partial y & 0 & 0 & 0 \\ 0 & \partial y & 0 & 0 \end{bmatrix} \qquad P_z = \begin{bmatrix} 0 & 0 & 0 & -\partial z \\ 0 & 0 & \partial z & 0 \\ 0 & -\partial z & 0 & 0 \\ \partial z & 0 & 0 & 0 \end{bmatrix}$$

$$(16.24)$$

Compared to the traditional $\mathbb{C}$ operators, (16.11), we see that the minus sign is in the conjugate and the i has become the quaternion $\{i, j, k\}$ in the form of the different matrices. We see that the function of the gamma matrices in the traditional Dirac equation is as the quaternion imaginary units $\{i, j, k\}$ but that we have achieved this by use of the matrices (16.24).

Substituting the differential operator matrices into the basic equation (16.23) and taking account of the two different non-commutative differentials (16.22) gives the quaternion version of the Dirac equation:

$$E_H Q = \frac{1}{2}(d_L + d_R)Q = mQ$$
$$(16.25)$$
$$B_H Q = \frac{1}{2}(d_L - d_R)Q = mQ$$

For a general quaternion field:

$$Q = \begin{bmatrix} \phi & A_x & A_y & A_z \\ -A_x & \phi & -A_z & A_y \\ -A_y & A_z & \phi & -A_x \\ -A_z & -A_y & A_x & \phi \end{bmatrix} \qquad (16.26)$$

The above, (16.25), quaternion Dirac equations are from:

$$d_L Q = \begin{bmatrix} \partial t & -\partial x & -\partial y & -\partial z \\ \partial x & \partial t & \partial z & -\partial y \\ \partial y & -\partial z & \partial t & \partial x \\ \partial z & \partial y & -\partial x & \partial t \end{bmatrix} \begin{bmatrix} \phi & A_x & A_y & A_z \\ -A_x & \phi & -A_z & A_y \\ -A_y & A_z & \phi & -A_x \\ -A_z & -A_y & A_x & \phi \end{bmatrix}$$

(16.27)

Which is (we give only the top row elements for ease of presentation):

$$d_L Q_{[1,1]} = \frac{\partial \phi}{\partial t} + \frac{\partial A_x}{\partial x} + \frac{\partial A_y}{\partial y} + \frac{\partial A_z}{\partial z}$$

$$d_L Q_{[1,2]} = \frac{\partial A_x}{\partial t} - \frac{\partial \phi}{\partial x} - \frac{\partial A_z}{\partial y} + \frac{\partial A_y}{\partial z}$$

$$d_L Q_{[1,3]} = \frac{\partial A_y}{\partial t} + \frac{\partial A_z}{\partial x} - \frac{\partial \phi}{\partial y} - \frac{\partial A_x}{\partial z}$$

(16.28)

$$d_L Q_{[1,4]} = \frac{\partial A_z}{\partial t} - \frac{\partial A_y}{\partial x} + \frac{\partial A_x}{\partial y} - \frac{\partial \phi}{\partial z}$$

and:

$$d_R Q = \begin{bmatrix} \phi & A_x & A_y & A_z \\ -A_x & \phi & -A_z & A_y \\ -A_y & A_z & \phi & -A_x \\ -A_z & -A_y & A_x & \phi \end{bmatrix} \begin{bmatrix} \partial t & -\partial x & -\partial y & -\partial z \\ \partial x & \partial t & \partial z & -\partial y \\ \partial y & -\partial z & \partial t & \partial x \\ \partial z & \partial y & -\partial x & \partial t \end{bmatrix}$$

(16.29)

Which is:

$$d_R Q_{[1,1]} = \frac{\partial \phi}{\partial t} + \frac{\partial A_x}{\partial x} + \frac{\partial A_y}{\partial y} + \frac{\partial A_z}{\partial z}$$

$$d_R Q_{[1,2]} = \frac{\partial A_x}{\partial t} - \frac{\partial \phi}{\partial x} + \frac{\partial A_z}{\partial y} - \frac{\partial A_y}{\partial z}$$

$$d_R Q_{[1,3]} = \frac{\partial A_y}{\partial t} - \frac{\partial A_z}{\partial x} - \frac{\partial \phi}{\partial y} + \frac{\partial A_x}{\partial z}$$

$$d_R Q_{[1,4]} = \frac{\partial A_z}{\partial t} + \frac{\partial A_y}{\partial x} - \frac{\partial A_x}{\partial y} - \frac{\partial \phi}{\partial z}$$

(16.30)

Using these, (16.28) & (16.30), we get the E-field to be:

$$E_\mathbb{H} Q_{[1,1]} = \frac{\partial \phi}{\partial t} + \frac{\partial A_x}{\partial x} + \frac{\partial A_y}{\partial y} + \frac{\partial A_z}{\partial z}$$

$$E_\mathbb{H} Q_{[1,2]} = \frac{\partial A_x}{\partial t} - \frac{\partial \phi}{\partial x}$$

$$E_\mathbb{H} Q_{[1,3]} = \frac{\partial A_y}{\partial t} - \frac{\partial \phi}{\partial y}$$

$$E_\mathbb{H} Q_{[1,4]} = \frac{\partial A_z}{\partial t} - \frac{\partial \phi}{\partial z}$$

(16.31)

These are the equations that we would normally take to define an electric field except for an arbitrary sign difference.

The B-field is:

$$B_\mathbb{H} Q_{[1,1]} = 0$$

$$B_\mathbb{H} Q_{[1,2]} = \frac{\partial A_y}{\partial z} - \frac{\partial A_z}{\partial y}$$

$$B_\mathbb{H} Q_{[1,3]} = \frac{\partial A_z}{\partial x} - \frac{\partial A_x}{\partial z}$$

$$B_\mathbb{H} Q_{[1,4]} = \frac{\partial A_x}{\partial y} - \frac{\partial A_y}{\partial x}$$

(16.32)

These are the equations that we would normally take to define a magnetic field to arbitrary signs.

The Dirac E-field equation is now the quaternion equation:

$$\frac{\partial \phi}{\partial t} + \frac{\partial A_x}{\partial x} + \frac{\partial A_y}{\partial y} + \frac{\partial A_z}{\partial z} - m\phi = 0$$

$$\frac{\partial A_x}{\partial t} - \frac{\partial \phi}{\partial x} - mA_x = 0$$

$$\frac{\partial A_y}{\partial t} - \frac{\partial \phi}{\partial y} - mA_y = 0 \qquad (16.33)$$

$$\frac{\partial A_z}{\partial t} - \frac{\partial \phi}{\partial z} - mA_z = 0$$

The Dirac B-field equation is now the quaternion equation:

$$m = 0$$

$$\frac{\partial A_y}{\partial z} - \frac{\partial A_z}{\partial y} - mA_x = 0$$

$$\frac{\partial A_z}{\partial x} - \frac{\partial A_x}{\partial z} - mA_y = 0 \qquad (16.34)$$

$$\frac{\partial A_x}{\partial y} - \frac{\partial A_y}{\partial x} - mA_z = 0$$

We take it that the Dirac E-field is the electron field.

The Dirac B-field is a massless particle. Further putting $m = 0$ into the other three equations in the Dirac B-field, (16.34), says that the magnetic moment of the particle of this field is zero. We take it that the Dirac B-field is the neutrino.

It would seem that, as the magnetic field is to the electric field within classical physics, the neutrino is to the electron in quantum electrodynamics. A nice result.

But I thought neutrinos had mass. We deal with this in the next chapter.

Note that (16.34) does not say there is no Dirac B-field. With zero mass, we have the Dirac B-field:

$$m = 0$$

$$\frac{\partial A_y}{\partial z} = \frac{\partial A_z}{\partial y}$$

$$\frac{\partial A_z}{\partial x} = \frac{\partial A_x}{\partial z} \qquad (16.35)$$

$$\frac{\partial A_x}{\partial y} = \frac{\partial A_y}{\partial x}$$

Other quaternion fields:

Suppose we have used a conjugate quaternion differential operator to act on a conjugate quaternion? Up to a few signs, we would have the same result.

How about an anti-quaternion? The E-field of an anti-quaternion is:

$$E_{\mathbb{H}_{Anti}} Q_{Anti[1,1]} = \frac{\partial \phi}{\partial t} + \frac{\partial A_x}{\partial x} + \frac{\partial A_y}{\partial y} + \frac{\partial A_z}{\partial z}$$

$$E_{\mathbb{H}_{Anti}} Q_{Anti[1,2]} = \frac{\partial A_x}{\partial t} - \frac{\partial \phi}{\partial x}$$

$$E_{\mathbb{H}_{Anti}} Q_{Anti[1,3]} = \frac{\partial A_y}{\partial t} - \frac{\partial \phi}{\partial y} \qquad (16.36)$$

$$E_{\mathbb{H}_{Anti}} Q_{Anti[1,4]} = \frac{\partial A_z}{\partial t} - \frac{\partial \phi}{\partial z}$$

This E-field of the anti-quaternions is the same as the E-field of the quaternion except that it is right-chiral rather than left-chiral. We have two electron fields which differ in only their chirality.

Note that these two electron fields should not be confused with the left-ward moving and right-ward moving electron fields we met circa (15.35). The electron fields in this chapter are left-chiral and right-chiral not left-ward moving anf right-ward moving.

The B-field of an anti-quaternion is:

$$B_{\mathbb{H}_{Anti}} Q_{Anti[1,1]} = 0$$

$$B_{\mathbb{H}_{Anti}} Q_{Anti[1,2]} = -\frac{\partial A_y}{\partial z} + \frac{\partial A_z}{\partial y}$$

$$B_{\mathbb{H}_{Anti}} Q_{Anti[1,3]} = -\frac{\partial A_z}{\partial x} + \frac{\partial A_x}{\partial z} \qquad (16.37)$$

$$B_{\mathbb{H}_{Anti}} Q_{Anti[1,4]} = -\frac{\partial A_x}{\partial y} + \frac{\partial A_y}{\partial x}$$

This is the reverse of the quaternion B-field, but it is also of the reverse chirality. It woud seem that the right-chiral neutrino-field is the reverse of the left-chiral neutrino-field whereas the right-chiral electron-field is the same as the left-chiral electron-field.

Such double reversal of the neutrino field explains why we see only neutrinos of one chirality. With zero mass, the anti-quaternion neutrino field is the same as the quaternion neutrino field, (16.35).

Summary:
We have followed the traditional derivation of the Dirac equation from the relativistic energy momentum relation except we have done this by a slightly different, and we think more efficient, route.

We have used the quaterion, $SU(2)$, differential operators rather than the traditional $\mathbb{C} \equiv U(1)$ differential operators, and this has led to two fields. We have taken these two fields to be the electron-field and the neutrino-field.

The maths drives us to take neutrinos to be massless. The maths drives us to have electrons of both chiralities but to have neutrinos of only one chirality.

The Neutrino Mass Problem Solved

Given the importance of the neutrino mas problem in modern quantum field theory, your considerate author thought that these next few paragraphs deserved a chapter of their own.

The neutrino mass problem stated:
Experiments, such as measuring the velocity of neutrinos, indicate that neutrinos are massless. Experiments such as measuring the flux of electron neutrinos from the sun indicate that neutrinos oscillate into one another and so have squared mass. We reproduce the probability calculation we presented earlier, (12.7), of an electron neutrino oscilating into muon neutrino:

$$\text{Prob}(v_e \to v_\mu) = \sin^2(2\theta)\sin^2\left(\frac{\left(m_e^2 - m_\mu^2\right)L}{4E}\right) \tag{17.1}$$

Thus, experiments indicate that neutrions have squared mass but zero mass. How can this be?

The neutrino mass problem solved:
The massless neutrino field derived above is a quaternion of the form, (16.32):

$$N = \begin{bmatrix} m & \dfrac{\partial A_y}{\partial z} - \dfrac{\partial A_z}{\partial y} & \dfrac{\partial A_z}{\partial x} - \dfrac{\partial A_x}{\partial z} & \dfrac{\partial A_x}{\partial y} - \dfrac{\partial A_y}{\partial x} \\[2.2ex] -\left(\dfrac{\partial A_y}{\partial z} - \dfrac{\partial A_z}{\partial y}\right) & m & -\left(\dfrac{\partial A_x}{\partial y} - \dfrac{\partial A_y}{\partial x}\right) & \dfrac{\partial A_z}{\partial x} - \dfrac{\partial A_x}{\partial z} \\[2.2ex] -\left(\dfrac{\partial A_z}{\partial x} - \dfrac{\partial A_x}{\partial z}\right) & \dfrac{\partial A_x}{\partial y} - \dfrac{\partial A_y}{\partial x} & m & -\left(\dfrac{\partial A_y}{\partial z} - \dfrac{\partial A_z}{\partial y}\right) \\[2.2ex] -\left(\dfrac{\partial A_x}{\partial y} - \dfrac{\partial A_y}{\partial x}\right) & -\left(\dfrac{\partial A_z}{\partial x} - \dfrac{\partial A_x}{\partial z}\right) & \dfrac{\partial A_y}{\partial z} - \dfrac{\partial A_z}{\partial y} & m \end{bmatrix}$$

$$m = 0$$

(17.2)

Let us square this quaternion. We multiply this quaternion by its conjugate to square it. We get a quaternion of the form:

$$NN^* = \begin{bmatrix} p_x^2 + p_y^2 + p_z^2 & 0 & 0 & 0 \\ 0 & p_x^2 + p_y^2 + p_z^2 & 0 & 0 \\ 0 & 0 & p_x^2 + p_y^2 + p_z^2 & 0 \\ 0 & 0 & 0 & p_x^2 + p_y^2 + p_z^2 \end{bmatrix}$$

(17.3)

wherein we have used $p_x = \dfrac{\partial A_y}{\partial z} - \dfrac{\partial A_z}{\partial y}$ and similar.

The important point is that a quaternion with zero real part has a square with a non-zero real part. We associate the mass of the neutrino with the real part of the quaternion. We see that a massless neutrino has a non-zero squared mass.

Thus, we can have both massless neutrinos and neutrino oscillation.

And Lepton number conservation?:

If neutrinos are such that they can oscillate from one type of neutrino to a different type of neutrino, then lepton number is not conserved within a single generation. Above, we have not found that neutrinos oscillate between generations; we have found that the squared neutrino field oscillates between generations. It would seem that we do not have separate generation lepton number conservation in respect of the squared neutrino fields. We are still allowed lepton number conservation within each separate generation of neutrinos.

A separate question is whether or not the squared neutrino field has lepton number; we do not know the answer.

Neutrino generations:

It is often said that the different generations of massive fermions differ from each other only by their mass. This is not quite true. If lepton number is conserved within each separate generation, then the different generations of massive fermions also differ by their generation lepton number.

Massless fermions (the neutrinos) can differ from each other by their generation lepton number.

Summary:

Although the neutrino field is massless, the neutrino field squared has mass. This allows neutrinos to travel at the speed of light and the different neutrino types to osscilate into each other.

Chapter 18

Electroweak Theory

We will not be looking at electroweak theory in any depth in this book because we would need to present the Higgs mechanism in preparation for presenting electroweak theory. The Higgs mechanism is drifting too far from the topic of this book, and so we will not present it to the reader. However, a book about the electron would be incomplete without a mention of electroweak theory. We therefore give a shallow presentation of this part of the electron story.

Introduction to electroweak theory:
Above, throughout this book, we have written of two different forces associated with the electron. One of these forces is the electromagnetic force which is the electric charge and magnetism associated with electrically charged particles. The other of these forces is the weak nuclear force which is a force associated with all leptons including the electrically neutral neutrinos. We have had, and continue to have, no interest in the gravitational force or the strong nuclear force.

The electromagnetic force and the weak nuclear force have been shown by Stephen Weinberg (1933-), Abdus Salam (1926-1996) and Sheldon Glashow (1932-) to be two different aspects of the same single force which is now called the electroweak interaction[95] or the electroweak force. The theory of the electroweak interaction is known as the Weinberg-Salam model. Stephen Weinberg, Abdus Salam and Sheldon Glashow shared the 1979 Nobel Prize for their parts in the development of this model. One of the remarkable aspects of the model is that, through the Higgs mechanism, Weinberg and

---

[95] Interaction is just a posh word for force.

Salam were able to predict the masses of the weak force bosons $W^{\pm}$ & Z^0 - see next paragraph.

The Weinberg-Salam model is based upon neutrinos being massless. The Weinberg-Salam model is a part of the Standard model.

Generators and Lie algebras:

Within quantum field theory, forces are transmitted between interacting paricles by bosons like the photon. The electromagnetic force is transmitted by the photon which is denoted by γ; the photon has no electric charge. The weak force is transmitted by three bosons known as the weak force bosons which are denoted by $\{Z^0, W^+, W^-\}$; the Z^0 boson has no electric charge; the W^+ has positive electric charge, and the W^- has negative electric charge. In case you were wondering, the W^+ is not the anti-particle of the W^-. Bosons are their own anti-particles, or perhaps they do not have anti-particles.

Within QFT, each boson is associated with the generator of a Lie group.[96] The idea is that each boson is associated, by Noether's theorem, with a rotational symmetry. Electroweak theory uses two Lie groups known as $U(1)$ and $SU(2)$. The Standard model of QFT is based on the three Lie groups $U(1)$ and $SU(2)$ and $SU(3)$. Electroweak theory is said to be based in $SU(2) \times U(1)$. Note that the cross does not have the same meaning here as it does in finite group theory. If fact, the cross is quite meaningless in Lie group theory.

The Lie group $U(1)$ is just rotation in the 2-dimensional complex plane, $\mathbb{C}$. Such rotation is expressed as the exponential of a 2×2 matrix:

$$U(1): \quad \exp\left(\begin{bmatrix} 0 & b \\ -b & 0 \end{bmatrix}\right) = \begin{bmatrix} \cos b & \sin b \\ -\sin b & \cos b \end{bmatrix} \qquad (18.1)$$

---

[96] See : Dennis Morris : Lie Groups and Lie Algebras

The matrix:

$$\begin{bmatrix} 0 & 1 \\ -1 & 0 \end{bmatrix} \tag{18.2}$$

is said to be the generator of the Lie group $U(1)$. Unfortunately, and confusingly, this generator is often written as the 1×1 matrix $[i]$ leading to:

$$\exp([ib]) = \cos b + i \sin b \tag{18.3}$$

which is a rotation matrix without the matrix. Your author sometimes thinks that such notation is deliberately used to confuse your author.

The Lie group $SU(2)$ is rotation in quaternion space. Unfortunately, the Lie algebraists seem to have not got it quite right in regard to the generators of $SU(2)$. Because the quaternions are a single division algebra, the quaternion rotation matrix is generated by taking the exponential of a single generator matrix. Lie algebraists, not knowing of rotations other than 2-dimensional rotations, separate out the three imaginary quaternion unit elements and call them the three $SU(2)$ generators. These $SU(2)$ generator matrices are:

$$\begin{bmatrix} 0 & 1 & 0 & 0 \\ -1 & 0 & 0 & 0 \\ 0 & 0 & 0 & -1 \\ 0 & 0 & 1 & 0 \end{bmatrix}, \quad \begin{bmatrix} 0 & 0 & 1 & 0 \\ 0 & 0 & 0 & 1 \\ -1 & 0 & 0 & 0 \\ 0 & -1 & 0 & 0 \end{bmatrix}, \quad \begin{bmatrix} 0 & 0 & 0 & 1 \\ 0 & 0 & -1 & 0 \\ 0 & 1 & 0 & 0 \\ -1 & 0 & 0 & 0 \end{bmatrix}$$

$$\tag{18.4}$$

To confuse your author further, these matrices are reduced to 2×2 matrices, multiplied by $i = \sqrt{-1}$, and written as:

$$\sigma_1 = \begin{bmatrix} 0 & 1 \\ 1 & 0 \end{bmatrix}, \quad \sigma_2 = \begin{bmatrix} 0 & -i \\ i & 0 \end{bmatrix}, \quad \sigma_2 = \begin{bmatrix} 1 & 0 \\ 0 & -1 \end{bmatrix} \tag{18.5}$$

These are called the Pauli matrices. The $SU(2)$ matrices can also be written as 3×3 matrices and as 4×4 matrices and as ..., and the Lie algebraists make a big fuss about these different sized representations of the $SU(2)$ generators.

When all is said and done, the photon is associated with the $U(1)$ generator, (18.2), and the three $\{W^\pm, Z^0\}$ bosons are associated with the $SU(2)$ generators (18.4) or (18.5).

Left chiral and right chiral electrons:
Electroweak theory takes an electron field within our 4-dimensional space-time to be a superposition of a left-chiral electron field and a right-chiral electron field.

What!left-chiral electrons and right-chiral electrons? I thought there was only one type of electron. There is only one type of electron in our 4-dimensional space-time. Within our 4-dimensional space-time, electrons have mass. Because electrons have mass in our 4-dimensional space-time, they move at less than the speed of light. Because electrons move at less than the speed of light, a moving electron can be observed by an observer moving slower than the electron to be spinning in a left-handed way; this observer might say that the electron is left-handed meaning that the angular momentum vector of the electron and the momentum vector of the electron both point in the same direction. However, the observer can accelerate to a velocity faster than the electron. An observer moving faster than the electron will see the electron moving backwards relative to herself, and she will see what is a left-handed electron to a slowly moving observer appear to her to be a right-handed electron to herself meaning that the electron's angular momentum vector is pointing in the direction opposite to the electron's momentum vector.

Within our 4-dimensional space-time, a right-handed electron is the same thing as a left-handed electron seen by a different observer – there is only one kind of electron. Great, that's what I thought.

However, if the electron is massless, it will move at the speed of light. But electrons are not massless. I know electrons are not massless, but, to construct electroweak theory, we have to imagine that electrons are massless. If an electron is moving at the speed of light, it cannot be overtaken by an observer, and so the electron is left-handed (or right-handed) to every observer. Thus, we have two types of massless electrons – left-handed and right-handed.

The reader might like to think of the massless electrons as existing outside of our 4-dimensional space-time and aquiring mass when they pop into our 4-dimensional space-time.

Electroweak theory is built upon the imagined concept of massless electrons.

Electroweak theory takes the massless Dirac lagrangian to be:

$$\mathcal{L}_{massless} = i\overline{\psi}\gamma^{\mu}\partial_{\mu}\psi \tag{18.6}$$

We take $\psi = \psi_L + \psi_R$, and this massless Dirac Lagrangian becomes:

$$\mathcal{L}_{massless} = i\left(\overline{\psi_L} + \overline{\psi_R}\right)\gamma^{\mu}\partial_{\mu}\left(\psi_L + \psi_R\right) \tag{18.7}$$

After a little manipulation[97], this becomes:

$$\mathcal{L}_{massless} = i\overline{\psi_L}\gamma^{\mu}\partial_{\mu}\psi_L + i\overline{\psi_R}\gamma^{\mu}\partial_{\mu}\psi_R \tag{18.8}$$

Electroweak theory then decides that the left-chiral electron field can be associated with a left-chiral neutrino field, but that the right-chiral electron field cannot be associated with a neutrino; this is presented as:

$$\psi_L = \begin{bmatrix} v_e \\ e_L^- \end{bmatrix}, \quad \psi_R = \begin{bmatrix} 0 \\ e_R^- \end{bmatrix} \tag{18.9}$$

We see here a chiral imbalance is deliberately introduced into electroweak theory.

---

[97] See: David McMahon Quantum Field Theory Demystified pg 212

There is more imbalance. Within electroweak theory there are two new charges called weak isospin, I^3, and weak hypercharge, Y. Left chiral fermions have:

$$\begin{aligned}
\nu_L \quad &: \quad I^3 = +\frac{1}{2}, \quad Y = -1 \\
e_L^- \quad &: \quad I^3 = -\frac{1}{2}, \quad Y = -1
\end{aligned} \tag{18.10}$$

The right chiral fermion has:

$$e_R^- \quad : \quad I^3 = 0, \qquad Y = -2 \tag{18.11}$$

The sum of these two new charges is the electric charge of the particle:

$$Q = I^3 + \frac{Y}{2} \tag{18.12}$$

This relation, (18.12), is called the Gell-Mann-Nishijima relation after Murray Gell-Mann (1929-) and Kazuhiko Nishijima (1926-2009).

A look back:

The reader might recall that above we found that the neutrino fields of the quaternions and the anti-quaternions share the same chirality with respect to the neutrino field, circa (16.37). Looking at the quaternions and the anti-quaternions, (12.11) & (12.12), we see an imbalance in the number of double clock-wise rotations to both clockwise and anti-clockwise rotations – the distribution of the minus signs within the matrices of the two quaternion algebras.

The unbalanced lepton lagrangian:

Mathematically, the imbalance in the left-chiral lepton fields, the left-chiral electron and the left-chiral neutrino, and the right-chiral lepton field, the right-chiral electron is expressed in the lepton part of the electroweak lagrangian which is:

$$\mathcal{L}_{lepton} = i\overline{\psi_R}\gamma^\mu \left(\partial_\mu + i\frac{g_B}{2}B_\mu\right)\psi_R + i\overline{\psi_L}\gamma^\mu \left(\partial_\mu + i\frac{g_B}{2}B_\mu + i\frac{g_W}{2}\vec{\tau}\cdot\vec{W}_\mu\right)\psi_L$$

(18.13)

wherein $\vec{W}_\mu$ is the three $SU(2)$ bosons, $\vec{\tau}$ is the three $SU(2)$ generators, g_W is a real number, called a coupling constant, which is a measure of the strength of the weak force, B_μ is the $U(1)$ generator, and g_B is a real number, a coupling constant, which is a measure of the strength of the electromagnetic force.

Within this lagrangian, (18.13), we see that the imbalance between left and right is the term $i\frac{g_W}{2}\vec{\tau}\cdot\vec{W}_\mu$.

The two terms, $\left(\partial_\mu + i\frac{g_B}{2}B_\mu\right)$ and $\left(\partial_\mu + i\frac{g_B}{2}B_\mu + i\frac{g_W}{2}\vec{\tau}\cdot\vec{W}_\mu\right)$ are called covariant derivatives. When differentiating, we have to adjust the result by adding the bits $i\frac{g_B}{2}B_\mu$ and $i\frac{g_B}{2}B_\mu + i\frac{g_W}{2}\vec{\tau}\cdot\vec{W}_\mu$ respectivly to the right fermion field and to the left fermion field.

The weak force bosons gain mass:
Without going into the details, applying the Higgs mechanism leads to the three $SU(2)$ bosons, $\{W^\pm, Z^0\}$ gaining mass but the $U(1)$ boson, B, remaining massless. The $U(1)$ boson is associated with the photon and thus is the boson of the electromagnetic interaction. Weinberg and Salam were able to predict the masses of the $SU(2)$ bosons to be:

$$W^\pm{}_{mass} \geq 38 \; GeV$$
$$Z^0{}_{mass} \geq 76 \; GeV$$

(18.14)

These masses were later verified by experiment.

Summary:

Electroweak theory is not pretty. The imbalance between left-chiral fermion fields and right-chiral fermion fields is fed into electroweak theory by hand to make the theory work. The concept of right-handed electrons and left-handed electrons is invented to make the theory work.

Electroweak theory relies upon the Higgs mechanism to give mass to the $SU(2)$ bosons. It is generally accepted, even given the discovery of the Higgs boson at the LHC, that the Higgs mechanism is contrived and not pretty.

Electrweak theory is an accepted part of the Standard model because it is able to predict the mass of the $SU(2)$ bosons, but most physicists would like something prettier.

Chapter 19

Concluding Remarks

We hope the reader has enjoyed this book. We hope the reader has a much deeper understanding of the fermions, the electron and the neutrino, than they had before reading this book. We hope we have presented the conventional QFT view of the electron fairly to the reader. However, we must admit that we prefer the quaternion view of the electron to the conventional pair of complex numbers view of the electron.

We feel that the quaternion view of the electron is far less obscure than the pair of complex numbers view. We feel that the association of the electroweak interaction and the Lie group $SU(2)$ is much more patent in the quaternion view of the electron. We feel that the quaternion Dirac equation is mathematically far more appealing than the conventional Dirac equation because the quaternion Dirac equation is written within the quaternion division algebra. We feel that the conventional view of the neutrino as being represented by the Dirac equation with zero mass is without merit and that representing the electron and the neutrino as the Dirac E-field and the Dirac B-field respectively is with merit.

We have at least presented to the reader a view of electron's intrinsic spin and why it is directionally quantitised which is different from the classical angular momentum conventionally associated with intrinsic spin. We feel we have presented to the reader much else besides. However, we must warn the reader that we gleaned these views by walking *"... a road less travelled ..."*, and that these views are outside of the conventional understanding of QFT.

This book is an early volume in what might become a whole new view of physics. Perhaps, fifty years from now, we will look back in awe at how simple the universe is and in amazement that the physicists and mathematicians of the 20th century were, in spite of tangling themselves with obscure and very obfuscating notation, able to come to the understanding we today call quantum electrodynamics and the Standard model. Your author hopes he has contributed to human understanding with this volume. Perhaps consequent to reading this book, the gentle reader will go on to deepen our world understanding to the benefit of all humankind.

It has been a pleasure to write for such a polite and understanding audience.

Dennis Morris

Port Mulgrave

May 2016

Other Books by the Same Author

The Naked Spinor – a Rewrite of Clifford Algebra

Spinors exist in Clifford algebras. In this book, we explore the nature of spinors.

Complex Numbers The Higher Dimensional Forms – Spinor Algebra

In this book, we explore the higher dimensional forms of complex numbers. These higher dimensional forms are closely connected to spinors.

Upon General Relativity

In this book, we see how 4-dimensional space-time, gravity, and electromagnetism emerge from the spinor algebras.

From Where Comes the Universe

This is a guide for the lay person to the physics of empty space.

Empty Space is Amazing Stuff – The Special Theory of Relativity

This book deduces the theory of special relativity from the group C_2. It gives a unique insight into the nature of the 2-dimensional space-time of special relativity.

The Nuts and Bolts of Quantum Mechanics

This is a gentle introduction to quantum mechanics for undergraduates.

Quaternions

This book pulls together the often separate properties of the quaternions. Non-commutative differentiation is covered as is non-commutative rotation and non-commutative inner products.

The Uniqueness of our Space-time

This book reports the finding that the only two geometric spaces within the finite groups are the two spaces that together form our universe. This is a startling finding. The nature of geometric space is explained alongside the nature on division algebra space.

Lie Groups and Lie Algebras

This book presents Lie theory from a diametrically different perspective to the usual presentation. This makes the subject much more intuitively obvious and easier to learn. Included is perhaps the clearest and simplest presentation of the true nature of the Lie group $SU(2)$ ever presented.

The Physics of Empty Space

This book presents a comprehensive understanding of empty space. The presence of 2-dimensional rotations in our 4-dimensional space-time is explained. Also included is a very gentle introduction to non-commutative differentiation. Classical electromagetism is deduced from the quaternions.

The Quaternion Dirac Equation

This small book (only 40 pages) presents the quaternion form of the Dirac equation. The neutrino mass problem is solved and we gain an explanation of why neutrinos are left-chiral. Much of the material in this book is drawn from 'The Electron'; this material is presented concisely and inexpensively for students already familiar with QFT.

An Essay on the Nature of Space-time

This small and inexpensive volume presents a view of the nature of empty space without the detailed mathematics. The expanding universe and dark energy is discussed.

Elementary Calculus from an Advanced Standpoint

This book rewrite the calculus of the complex numbers in a way that covers all division algebras and makes all continuous complex functions differentiable and integrable. Non-commutative differentiation is covered. Gauge covariant differentiation is covered as is the covariant derivative of general relativity.

Even Mathematicians and Physicists make Mistakes

This book points out what seems to be several important errors of modern physics and modern mathematics. Errors like the misunderstanding of rotation, the failure to teach the higher dimensional complex numbers in most universities, and the mathematical inconsistency of the Dirac equation and some casual errors are discussed. These errors are set in their historical circumstances and there is discussion about why they happened and the consequences of their happening. There is also an interesting chapter on the nature of mathematical proof within our society, and several famous proofs are discussed (without the details).

Finite Groups – A Simple Introduction

This book introduces the reader to finite group theory. Many introductory books on finite groups bury the reader in geometrical examples or in other types of groups and lose the central nature of a finite group. This book sticks firmly with the permutation nature of finite groups and elucidates that nature by the extensive use of permutation matrices. Permutation matrices simplify the subject considerably. This book is probably unique in its use of permutation matrices and therefore unique in its simplicity.

The Left-handed Spinor

This book covers the left-handed parts of mathematics which we call the chiral algebras. These algebras have CP invariance, violation of parity, and many other aspects which makes them relevant to theoretical physics. It is quite a revelation to discover that mathematics is left-handed.

Non-commutative Differentiation and the Commutator

(The Search for the Fermion Content of the Universe)

This book develops the theory of non-commutative differentiation from the fundamentals of algebra. We see what an algebraic operation (addition, multiplication) really is, and we discover that the commutator is a third fundamental algebraic operation within some division algebras. This leads to the first part of the derivation of the fermion content of the universe.

Bibliography

The Standard Model	Cliff Burgess anf Guy Moore
The Ideas of Particle Physics	Coughan, Dodd, & Gripaios
Lectures on Physics	Feynmann, Leighton, & Sands
An Introduction to Quantum Physics	French & Taylor
Lie Algebras in Particle Physics	Howard Georgi
Introductory Quantum Theory	R. C. Greenhow
Supersymmetry DeMystified	Patrick Labelle
Quantum Field Theory DeMystified	David McMahon
Quantum Mechanics DeMystified	David McMahon
Quarks & Leptons	Halzen & Martin
Clifford Algebras and Spinors	Pertti Lounesto
Quantum Mechanics and the Particles of Nature	Sudbury
Quantum Field Theory in a Nutshell	A Zee

Index

D

E

F